# Differential Equati[...] For Dummies® [...] Sheet



## Classifying Differential Equations by Order

**First order differential equations** involve derivatives of the first order, such as in this example:

$$\frac{dy}{dx} = 5x$$

**Second order differential equations** involve derivatives of the second order, such as in these examples:

$$\frac{d^2y}{dx^2} = \sin(x) \qquad \frac{d^2y}{dx^2} + \frac{dy}{dx} = \cos(x)$$

**Higher order differential equations** are those involving derivatives higher than the second order (big surprise on that clever name!). Differential equations of all orders can use the $y'$ notation, like this:

$$\frac{dy}{dx} = y' = f(x) \qquad \frac{d^3y}{dx^3} = y''' - f(x) \qquad \frac{d^5y}{dx^5} = y^{(5)} = f(x)$$

$$\frac{d^2y}{dx^2} = y'' = f(x) \qquad \frac{d^4y}{dx^4} = y^{(4)} = f(x)$$

## Distinguishing among Linear, Separable, and Exact Differential Equations

**Linear differential equations,** which I introduce in Chapter 2, involve only derivatives of $y$ and terms of $y$ to the first power, not raised to any higher power. (***Note:*** This is the power the derivative is *raised to,* not the *order* of the derivative.) For example, this is a linear differential equation because it contains only derivatives raised to the first power:

$$\frac{d^4y}{dx^4} + \frac{d^3y}{dx^2} = f(x)$$

**Separable differential equations,** which I cover in Chapter 3, can be written so that all terms in $x$ and all terms in $y$ appear on opposite sides of the equation. Here's an example:

$$\frac{dy}{dx} = \frac{x^2}{2 - y^2}$$

which can be written like this with a little reshuffling:

$$(2 - y^2)dy = x^2dx$$

**Exact differential equations** are those where you can find a function whose partial derivatives correspond to the terms in a given differential equation. Chapter 4 has the information you need to know on these equations.

*For Dummies: Bestselling Book Series for Beginners*

# Differential Equations For Dummies®

Cheat Sheet

## Defining Homogeneous and Nonhomogeneous Differential Equations

In Part II you find my discussion on homogeneous and nonhomogeneous equations. What are those, you ask? **Homogeneous differential equations** involve only derivatives of *y* and terms involving *y*, and they're set to 0, such as in this equation:

$$\frac{d^4y}{dx^4} + x\frac{d^2y}{dx^2} + y^2 = 0$$

**Nonhomogeneous differential equations** are the same as homogeneous differential equations, except they can have terms involving only *x* (and constants) on the right side, such as in this equation:

$$\frac{d^4y}{dx^4} + x\frac{d^2y}{dx^2} + y^2 = 6x + 3$$

You also can write nonhomogeneous differential equations in this format: $y'' + p(x)y' + q(x)y = g(x)$. The general solution of this nonhomogeneous differential equation is the following:

$$y = c_1y_1(x) + c_2y_2(x) + y_p(x)$$

In this solution, $c_1y_1(x) + c_2y_2(x)$ is the general solution of the corresponding homogeneous differential equation:

$$y'' + p(x)y' + q(x)y = 0$$

And $y_p(x)$ is a specific solution to the nonhomogeneous equation.

## Using the Method of Undetermined Coefficients

Suppose you face the following nonhomogeneous differential equation:

$$y'' + p(x)y' + q(x)y = g(x)$$

The **method of undetermined coefficients** notes that when you find a candidate solution, *y*, and plug it into the left-hand side of the equation, you end up with $g(x)$. Because $g(x)$ is only a function of *x*, you can often guess the form of $y_p(x)$, up to arbitrary coefficients, and then solve for those coefficients by plugging $y_p(x)$ into the differential equation.

This method works because you're dealing only with $g(x)$, and the form of $g(x)$ can often tell you what a particular solution looks like. See Chapter 6 for full details.

# Differential Equations FOR DUMMIES®

# Differential Equations FOR DUMMIES®

by Steven Holzner, PhD

WILEY

Wiley Publishing, Inc.

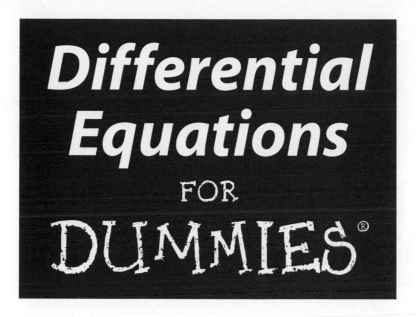

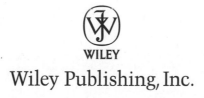

**Differential Equations For Dummies®**
Published by
**Wiley Publishing, Inc.**
111 River St.
Hoboken, NJ 07030-5774
www.wiley.com



For general information on our other products and services, please contact our Customer Care Department within the U.S. at 800-762-2974, outside the U.S. at 317-572-3993, or fax 317-572-4002.

For technical support, please visit www.wiley.com/techsupport.

Wiley also publishes its books in a variety of electronic formats. Some content that appears in print may not be available in electronic books.

Library of Congress Control Number: 2008925781

ISBN: 978-0-470-17814-0

Manufactured in the United States of America

10 9 8 7 6 5 4 3 2 1

WILEY

# About the Author

**Steven Holzner** is an award-winning author of science, math, and technical books. He got his training in differential equations at MIT and at Cornell University, where he got his PhD. He has been on the faculty at both MIT and Cornell University, and has written such bestsellers as *Physics For Dummies* and *Physics Workbook For Dummies.*

# Dedication

To Nancy, always and forever.

# Author's Acknowledgments

The book you hold in your hands is the work of many people. I'd especially like to thank Tracy Boggier, Georgette Beatty, Jessica Smith, technical reviewer Jamie Song, PhD, and the folks in Composition Services who put the book together so beautifully.

## Publisher's Acknowledgments

We're proud of this book; please send us your comments through our Dummies online registration form located at www.dummies.com/register/.

Some of the people who helped bring this book to market include the following:

***Acquisitions, Editorial, and Media Development***

**Project Editor:** Georgette Beatty

**Acquisitions Editor:** Tracy Boggier

**Copy Editor:** Jessica Smith

**Editorial Program Coordinator:** Erin Calligan Mooney

**Technical Editor:** Jamie Song, PhD

**Editorial Manager:** Michelle Hacker

**Editorial Assistants:** Joe Niesen, Leeann Harney

**Cartoons:** Rich Tennant (www.the5thwave.com)

***Composition Services***

**Project Coordinator:** Erin Smith

**Layout and Graphics:** Carrie A. Cesavice, Stephanie D. Jumper

**Proofreaders:** Caitie Kelly, Linda D. Morris

**Indexer:** Broccoli Information Management

---

**Publishing and Editorial for Consumer Dummies**

**Diane Graves Steele,** Vice President and Publisher, Consumer Dummies

**Joyce Pepple,** Acquisitions Director, Consumer Dummies

**Kristin A. Cocks,** Product Development Director, Consumer Dummies

**Michael Spring,** Vice President and Publisher, Travel

**Kelly Regan,** Editorial Director, Travel

**Publishing for Technology Dummies**

**Andy Cummings,** Vice President and Publisher, Dummies Technology/General User

**Composition Services**

**Gerry Fahey,** Vice President of Production Services

**Debbie Stailey,** Director of Composition Services

# Contents at a Glance

# Table of Contents

## Chapter 2: Looking at Linear First Order Differential Equations . . . . .23



## Chapter 3: Sorting Out Separable First Order Differential Equations . . . . . . . . . . . . . . . . . . . . . . . . . . . . . . . . . . . . . .41

# Introduction

For too many people who study differential equations, their only exposure to this amazingly rich and rewarding field of mathematics is through a textbook that lands with an 800-page whump on their desk. And what follows is a weary struggle as the reader tries to scale the impenetrable fortress of the massive tome.

Has no one ever thought to write a book on differential equations from the *reader's* point of view? Yes indeed — that's where this book comes in.

## About This Book

*Differential Equations For Dummies* is all about differential equations from *your* point of view. I've watched many people struggle with differential equations the standard way, and most of them share one common feeling: Confusion as to what they did to deserve such torture.

This book is different; rather than being written from the professor's point of view, it has been written from the reader's point of view. This book was designed to be crammed full of the good stuff, and *only* the good stuff. No extra filler has been added; and that means the issues aren't clouded. In this book, you discover ways that professors and instructors make solving problems simple.

You can leaf through this book as you like. In other words, it isn't important that you read it from beginning to end. Like other *For Dummies* books, this one has been designed to let you skip around as much as possible — this is your book, and now differential equations are your oyster.

## Conventions Used in This Book

Some books have a dozen confusing conventions that you need to know before you can even start reading. Not this one. Here are the few simple conventions that I include to help you navigate this book:

- *Italics* indicate definitions and emphasize certain words. As is customary in the math world, I also use italics to highlight variables.
- **Boldfaced** text highlights important theorems, matrices (arrays of numbers), keywords in bulleted lists, and actions to take in numbered steps.
- `Monofont` points out Web addresses.

When this book was printed, some Web addresses may have needed to break across two lines of text. If that happens, rest assured that I haven't put in any extra characters (such as hyphens) to indicate the break. So when using one of these Web addresses, type in exactly what you see in this book, pretending as though the line break doesn't exist.

## What You're Not to Read

Throughout this book, I share bits of information that may be interesting to you but not crucial to your understanding of an aspect of differential equations. You'll see this information either placed in a sidebar (a shaded gray box) or marked with a Technical Stuff icon. I won't be offended if you skip any of this text — really!

## Foolish Assumptions

This book assumes that you have no experience solving differential equations. Maybe you're a college student freshly enrolled in a class on differential equations, and you need a little extra help wrapping your brain around them. Or perhaps you're a student studying physics, chemistry, biology, economics, or engineering, and you quickly need to get a handle on differential equations to better understand your subject area.

Any study of differential equations takes as its starting point a knowledge of calculus. So I wrote this book with the assumption in mind that you know how to take basic derivatives and how to integrate. If you're totally at sea with these tasks, pick up a copy of *Calculus For Dummies* by Mark Ryan (Wiley) before you pick up this book.

## How This Book Is Organized

The world of differential equations is, well, big. And to handle it, I break that world down into different parts. Here are the various parts you see in this book.

## Part I: Focusing on First Order Differential Equations

I start this book with first order differential equations — that is, differential equations that involve derivatives to the first power. You see how to work with linear first order differential equations (*linear* means that the derivatives aren't squared, cubed, or anything like that). You also discover how to work with separable first order differential equations, which can be separated so that only terms in *y* appear on one side, and only terms in *x* (and constants) appear on the other. And, finally, in this part, you figure out how to handle exact differential equations. With this type of equation you try to find a function whose partial derivatives correspond to the terms in a differential equation (which makes solving the equation much easier).

## Part II: Surveying Second and Higher Order Differential Equations

In this part, I take things to a whole new level as I show you how to deal with second order and higher order differential equations. I divide equations into two main types: linear homogeneous equations and linear nonhomogeneous equations. You also find out that a whole new array of dazzling techniques can be used here, such as the method of undetermined coefficients and the method of variation of parameters.

## Part III: The Power Stuff: Advanced Techniques

Some differential equations are tougher than others, and in this part, I bring out the big guns. You see heavy-duty techniques like Laplace transforms and series solutions, and you start working with systems of differential equations. You also figure out how to use numerical methods to solve differential equations. These are methods of last resort, but they rarely fail.

## Part IV: The Part of Tens

You see the Part of Tens in all *For Dummies* books. This part is made up of fast-paced lists of ten items each; in this book, you find ten online differential equation tutorials and ten top online tools for solving differential equations.

# Icons Used in This Book

You can find several icons in the margins of this book, and here's what they mean:

This icon marks something to remember, such as a law of differential equations or a particularly juicy equation.

The text next to this icon is technical, insider stuff. You don't have to read it if you don't want to, but if you want to become a differential equations pro (and who doesn't?), take a look.

This icon alerts you to helpful hints in solving differential equations. If you're looking for shortcuts, search for this icon.

When you see this icon, watch out! It indicates something particularly tough to keep an eye out for.

# Where to Go from Here

You're ready to jump into Chapter 1. However, you don't have to start there if you don't want to; you can jump in anywhere you like — this book was written to allow you to do just that. But if you want to get the full story on differential equations from the beginning, jump into Chapter 1 first — that's where all the action starts.

# Part I
# Focusing on First Order Differential Equations

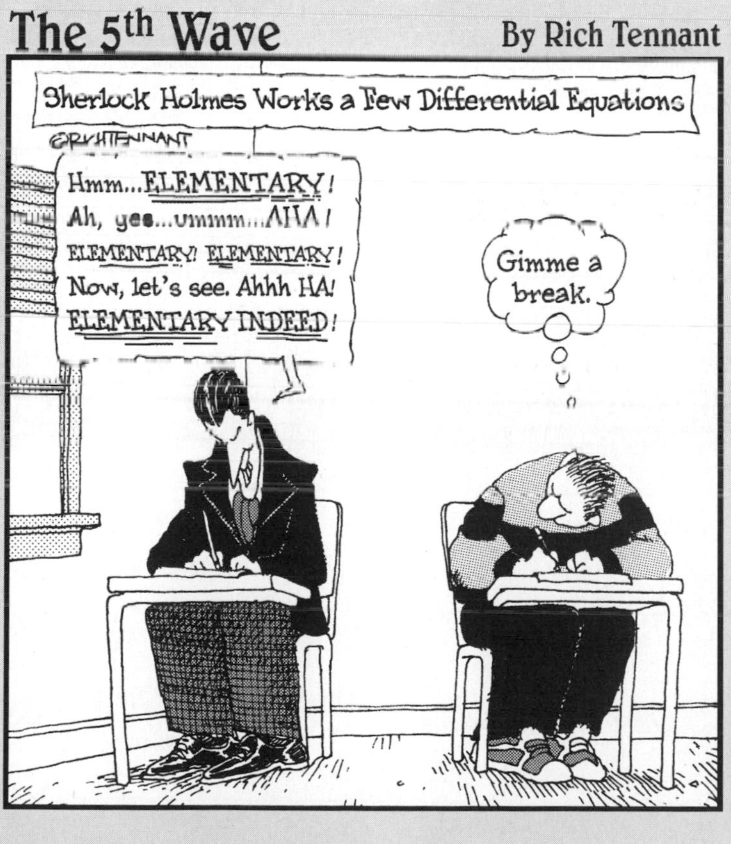

## In this part . . .

In this part, I welcome you to the world of differential equations and start you off easy with linear first order differential equations. With first order equations, you have first order derivatives that are raised to the first power, not squared or raised to any higher power. I also show you how to work with separable first order differential equations, which are those equations that can be separated so that terms in $y$ appear on one side and terms in $x$ (and constants) appear on the other. Finally, I introduce exact differential equations and Euler's method.

## Chapter 1

# Welcome to the World of Differential Equations

### *In This Chapter*

- Breaking into the basics of differential equations
- Getting the scoop on derivatives
- Checking out direction fields
- Putting differential equations into different categories
- Distinguishing among different orders of differential equations
- Surveying some advanced methods

It's a tense moment in the physics lab. The international team of high-powered physicists has attached a weight to a spring, and the weight is bouncing up and down.

"What's happening?" the physicists cry. "We have to understand this in terms of math! We need a formula to describe the motion of the weight!"

You, the renowned Differential Equations Expert, enter the conversation calmly. "No problem," you say. "I can derive a formula for you that will describe the motion you're seeing. But it's going to cost you."

The physicists look worried. "How much?" they ask, checking their grants and funding sources. You tell them.

"Okay, anything," they cry. "Just give us a formula."

You take out your clipboard and start writing.

"What's that?" one of the physicists asks, pointing at your calculations.

"That," you say, "is a differential equation. Now all I have to do is to solve it, and you'll have your formula." The physicists watch intently as you do your math at lightning speed.

"I've got it," you announce. "Your formula is $y = 10 \sin(5t)$, where $y$ is the weight's vertical position, and $t$ is time, measured in seconds."

"Wow," the physicists cry, "all that just from solving a differential equation?"

"Yep," you say, "now pay up."

Well, you're probably not a renowned differential equations expert — not yet, at least! But with the help of this book, you very well may become one. In this chapter, I give you the basics to get started with differential equations, such as derivatives, direction fields, and equation classifications.

# The Essence of Differential Equations

REMEMBER

In essence, differential equations involve *derivatives,* which specify how a quantity changes; by solving the differential equation, you get a formula for the quantity itself that doesn't involve derivatives.

Because derivatives are essential to differential equations, I take the time in the next section to get you up to speed on them. (If you're already an expert on derivatives, feel free to skip the next section.) In this section, however, I take a look at a qualitative example, just to get things started in an easily digestible way.

Say that you're a long-time shopper at your local grocery store, and you've noticed prices have been increasing with time. Here's the table you've been writing down, tracking the price of a jar of peanut butter:

| *Month* | *Price* |
|---|---|
| 1 | $2.40 |
| 2 | $2.50 |
| 3 | $2.60 |
| 4 | $2.70 |
| 5 | $2.80 |
| 6 | $2.90 |

Looks like prices have been going up steadily, as you can see in the graph of the prices in Figure 1-1. With that large of a price hike, what's the price of peanut butter going to be a year from now?

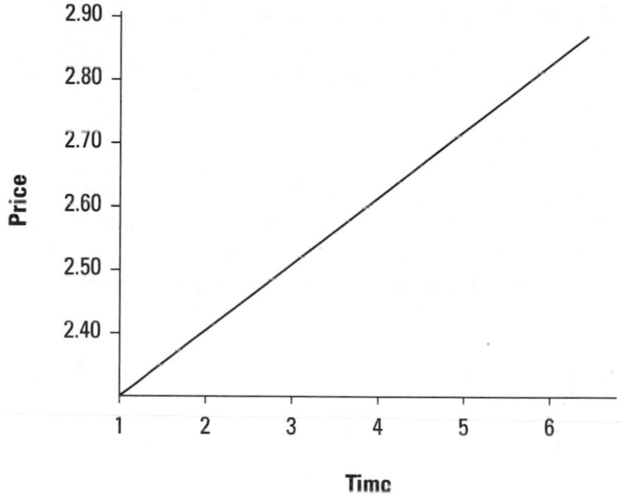


**Figure 1-1:** The price of peanut butter by month.

You know that the slope of a line is $\Delta y/\Delta x$ (that is, the change in $y$ divided by the change in $x$). Here, you use the symbols $\Delta p$ for the change in price and $\Delta t$ for the change in time. So the slope of the line in Figure 1-1 is $\Delta p/\Delta t$.

Because the price of peanut butter is going up 10 cents every month, you know that the slope of the line in Figure 1-1 is:

$$\frac{\Delta p}{\Delta t} = 10¢/\text{month}$$

The slope of a line is a constant, indicating its rate of change. The derivative of a quantity also gives its rate of change at any one point, so you can think of the derivative as the slope at a particular point. Because the rate of change of a line is constant, you can write:

$$\frac{dp}{dt} = \frac{\Delta p}{\Delta t} = 10¢/\text{month}$$

TIP

In this case, *dp*/*dt* is the derivative of the price of peanut butter with respect to time. (When you see the *d* symbol, you know it's a derivative.)

And so you get this differential equation:

$$\frac{dp}{dt} = 10¢/\text{month}$$

The previous equation is a differential equation because it's an equation that involves a derivative, in this case, $dp/dt$. It's a pretty simple differential equation, and you can solve for price as a function of time like this:

$$p = 10t + c$$

In this equation, $p$ is price (measured in cents), $t$ is time (measured in months), and $c$ is an arbitrary constant that you use to match the initial conditions of the problem. (You need a constant, $c$, because when you take the derivative of $10t + c$, you just get 10, so you can't tell whether there's a constant that should be added to $10t$ — matching the initial conditions will tell you.)

The missing link is the value of $c$, so just plug in the numbers you have for price and time to solve for it. For example, the cost of peanut butter in month 1 is \$2.40, so you can solve for $c$ by plugging in 1 for $t$ and \$2.40 for $p$ (240 cents), giving you:

$$240 = 10 + c$$

By solving this equation, you calculate that $c = 230$, so the solution to your differential equation is:

$$p = 10t + 230$$

And that's your solution — that's the price of peanut butter by month. You started with a differential equation, which gave the rate of change in the price of peanut butter, and then you solved that differential equation to get the price as a function of time, $p = 10t + 230$.

Want to see the solution to your differential equation in action? Go for it! Find out what the price of peanut butter is going to be in month 12. Now that you have your equation, it's easy enough to figure out:

$$p = 10t + 230$$

$$10(12) + 230 = 350$$

As you can see, in month 12, peanut butter is going to cost a steep \$3.50, which you were able to figure out because you knew the *rate* at which the price was increasing. This is how any typical differential equation may work: You have a differential equation for the rate at which some quantity changes (in this case, price), and then you solve the differential equation to get another equation, which in this case related price to time.

Note that when you substitute the solution ($p = 10t + 230$) into the differential equation, $dp/dt$ indeed gives you 10 cents per month, as it should.

# Derivatives: The Foundation of Differential Equations

As I mention in the previous section, a derivative simply specifies the rate at which a quantity changes. In math terms, the derivative of a function $f(x)$, which is depicted as $df(x)/dx$, or more commonly in this book, as $f'(x)$, indicates how $f(x)$ is changing at any value of $x$. The function $f(x)$ has to be continuous at a particular point for the derivative to exist at that point.

Take a closer look at this concept. The amount $f(x)$ changes in a small distance along the $x$ axis $\Delta x$ is:

$$f(x + \Delta x) - f(x)$$

The rate at which $f(x)$ changes over the change $\Delta x$ is:

$$\frac{f(x+\Delta x)-f(x)}{\Delta x}$$

So far so good. Now to get the derivative $dy/dx$, where $y = f(x)$, you must let $\Delta x$ get very small, approaching zero. You can do that with a *limiting expression*, which you can evaluate as $\Delta x$ goes to zero. In this case, the limiting expression is:

$$\frac{dy}{dx} = \lim_{\Delta x \to 0} \frac{f(x+\Delta x)-f(x)}{\Delta x}$$

In other words, the derivative of $f(x)$ is the amount $f(x)$ changes in $\Delta x$, divided by $\Delta x$, as $\Delta x$ goes to zero.

I take a look at some common derivatives in the following sections; you'll see these derivatives throughout this book.

## Derivatives that are constants

The first type of derivative you'll encounter is when $f(x)$ equals a constant, $c$. If $f(x) = c$, then $f(x + \Delta x) = c$ also, and $f(x + \Delta x) - f(x) = 0$ (because all these amounts are actually the same), so $df(x)/dx = 0$. Therefore:

$$f(x) = c \qquad \frac{df(x)}{dx} = 0$$

How about when $f(x) = cx$, where $c$ is a constant? In this case, $f(x) = cx$, and $f(x + \Delta x) = cx + c\,\Delta x$.

So $f(x + \Delta x) - f(x) = c\,\Delta x$ and $(f(x + \Delta x) - f(x))/\Delta x = c$. Therefore:

$$f(x) = cx \qquad \frac{df(x)}{dx} = c$$

## Derivatives that are powers

Another type of derivative that pops up is one that includes raising $x$ to the power $n$. Derivatives with powers work like this:

$$f(x) = x^n \qquad \frac{df(x)}{dx} = n\,x^{n-1}$$

Raising $e$ to a certain power is always popular when working with differential equations ($e$ is the natural logarithm base, $e = 2.7128\ldots$, and $a$ is a constant):

$$f(x) = e^{ax} \qquad \frac{df(x)}{dx} = a\,e^{ax}$$

And there's also the inverse of $e^a$, which is the natural log, which works like this:

$$f(x) = \ln(x) \qquad \frac{df(x)}{dx} = \frac{1}{x}$$

## Derivatives involving trigonometry

Now for some trigonometry, starting with the derivative of $\sin(x)$:

$$f(x) = \sin(x) \qquad \frac{df(x)}{dx} = \cos(x)$$

And here's the derivative of $\cos(x)$:

$$f(x) = \cos(x) \qquad \frac{df(x)}{dx} = -\sin(x)$$

## Derivatives involving multiple functions

The derivative of the sum (or difference) of two functions is equal to the sum (or difference) of the derivatives of the functions (that's easy enough to remember!):

$$f(x) = a(x) \pm b(x) \qquad \frac{df(x)}{dx} = \frac{d\,a(x)}{dx} \pm \frac{d\,b(x)}{dx}$$

The derivative of the product of two functions is equal to the first function times the derivative of the second, plus the second function times the derivative of the first. For example:

$$f(x) = a(x)b(x) \qquad \frac{df(x)}{dx} = a(x)\frac{d\,b(x)}{dx} + b(x)\frac{d\,a(x)}{dx}$$

How about the derivative of the quotient of two functions? That derivative is equal to the function in the denominator times the derivative of the function in the numerator, minus the function in the numerator times the derivative of the function in the denominator, all divided by the square of the function in the denominator:

$$f(x) = \frac{a(x)}{b(x)} \qquad \frac{df(x)}{dx} = \frac{b(x)\frac{d\,a(x)}{dx} - a(x)\frac{d\,b(x)}{dx}}{b(x)^2}$$

# Seeing the Big Picture with Direction Fields

It's all too easy to get caught in the math details of a differential equation, thereby losing any idea of the bigger picture. One useful tool for getting an overview of differential equations is a direction field, which I discuss in more detail in Chapter 2. *Direction fields* are great for getting a handle on differential equations of the following form:

$$\frac{dy}{dx} = f(x,y)$$

REMEMBER

The previous equation gives the slope of the equation $y = f(x)$ at any point $x$. A direction field can help you visualize such an equation without actually having to solve for the solution. That field is a two-dimensional graph consisting of many, sometimes hundreds, of short line segments, showing the slope — that is, the value of the derivative — at multiple points. In the following sections, I walk you through the process of plotting and understanding direction fields.

## Plotting a direction field

Here's an example to give you an idea of what a direction field looks like. A body falling through air experiences this force:

$$F = mg - \gamma v$$

In this equation, $F$ is the net force on the object, $m$ is the object's mass, $g$ is the acceleration due to gravity ($g = 9.8$ meters/sec$^2$ near the Earth's surface), $\gamma$ is the *drag coefficient* (which adds the effect of air friction and is measured in newtons sec/meter), and $v$ is the speed of the object as it plummets through the air.

If you're familiar with physics, consider Newton's second law. It says that $F = ma$, where $F$ is the net force acting on an object, $m$ is its mass, and $a$ is its acceleration. But the object's acceleration is also $dv/dt$, the derivative of the object's speed with respect to time (that is, the rate of change of the object's speed). Putting all this together gives you:

$$F = ma = m\frac{dv}{dt} = mg - \gamma v$$

Now you're back in differential equation territory, with this differential equation for speed as a function of time:

$$\frac{dv}{dt} = g - \frac{\gamma}{m}v$$

Now you can get specific by plugging in some numbers. The acceleration due to gravity, $g$, is 9.8 meters/sec$^2$ near the Earth's surface, and let's say that the drag coefficient is 1.0 newtons sec/meter and the object has a mass of 4.0 kilograms. Here's what you'd get:

$$\frac{dv}{dt} = 9.8 - \frac{v}{4}$$

REMEMBER

To get a handle on this equation without attempting to solve it, you can plot it as a direction field. To do so you create a two-dimensional plot and add dozens of short line segments that give the slope at those locations (you can do this by hand or with software). The direction field for this equation appears in Figure 1-2. As you can see in the figure, there are dozens of short lines in the graph, each of which give the slope of the *solution* at that point. The vertical axis is $v$, and the horizontal axis is $t$.

TIP

Because the slope of the solution function at any one point doesn't depend on $t$, the slopes along any horizontal line are the same.

## Connecting slopes into an integral curve

REMEMBER

You can get a visual handle on what's happening with the solutions to a differential equation by looking at its direction field. How? All those slanted line segments give you the solutions of the differential equations — all you have to do is draw lines connecting the slopes. One such solution appears in Figure 1-3. A solution like the one in the figure is called an *integral curve* of the differential equation.

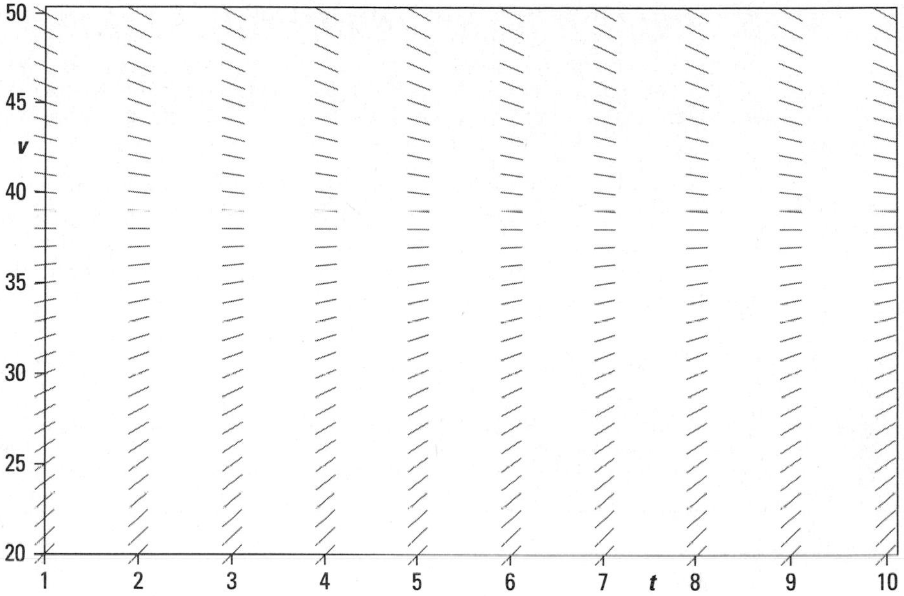


**Figure 1-2:** A direction field.

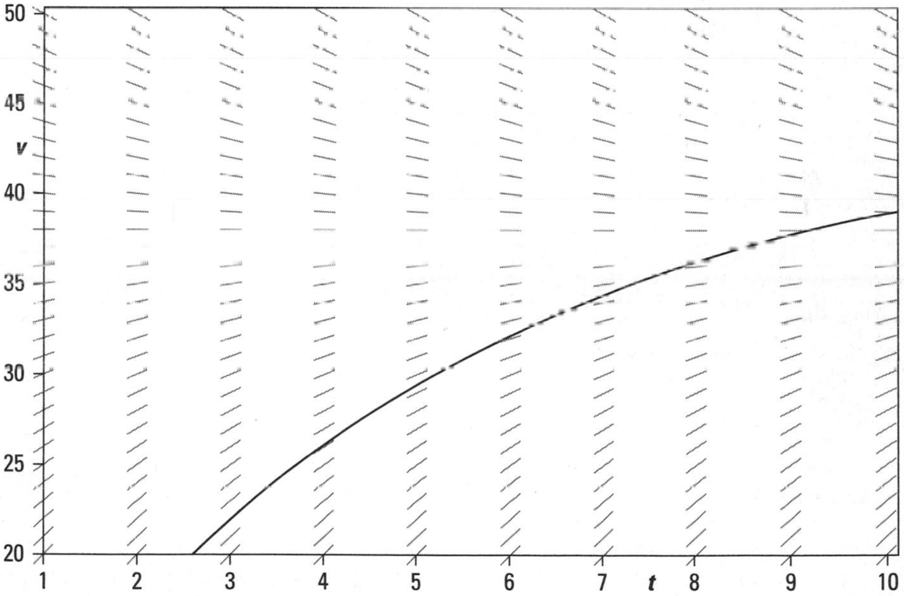


**Figure 1-3:** A solution in a direction field.

## Recognizing the equilibrium value

As you can see from Figure 1-3, there are many solutions to the equation that you're trying to solve. As it happens, the actual solution to that differential equation is:

$$v = 39.2 + ce^{-t/4}$$

In the previous solution, $c$ is an arbitrary constant that can take any value. That means there are an infinite number of solutions to the differential equation.

REMEMBER

But you don't have to know that solution to determine what the solutions behave like. You can tell just by looking at the direction field that all solutions tend toward a particular value, called the *equilibrium value.* For instance, you can see from the direction field graph in Figure 1-3 that the equilibrium value is 39.2. You also can see that equilibrium value in Figure 1-4.

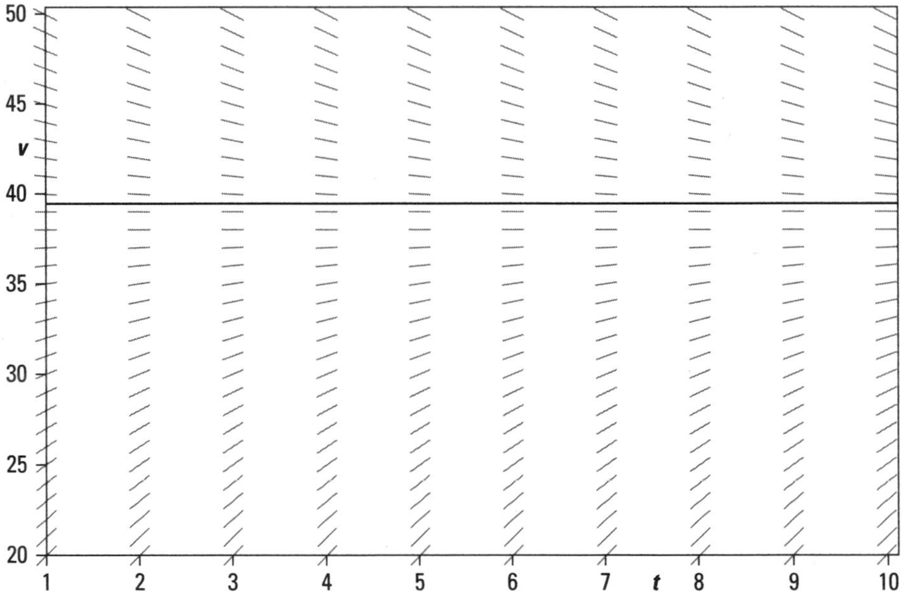


**Figure 1-4:** An equilibrium value in a direction field.

# Classifying Differential Equations

Tons of differential equations exist in Math and Science Land, and the way you tackle them differs by type. As a result, there are several *classifications* that you can put differential equations into. I explain them in the following sections.

## Classifying equations by order

REMEMBER

The most common classification of differential equations is based on *order*. The order of a differential equation simply is the order of its highest derivative. For example, check out the following, which is a first order differential equation:

$$\frac{dy}{dx} = 5x$$

Here's an example of a second order differential equation:

$$\frac{d^2y}{dx^2} + \frac{dy}{dx} = 19x + 4$$

And so on, up to order $n$:

$$9\frac{d^ny}{dx^n} - 16\frac{d^{n-1}y}{dx^{n-1}} + \cdots + 14\frac{d^2y}{dx^2} + 12\frac{dy}{dx} - 19x + 4 = 0$$

As you might imagine, first order differential equations are usually the most easily managed, followed by second order equations, and so on. I discuss first order, second order, and higher order differential equations in a bit more detail later in this chapter.

## Classifying ordinary versus partial equations

You can also classify differential equations as *ordinary* or *partial*. This classification depends on whether you have only ordinary derivatives involved or only partial derivatives.

REMEMBER

An *ordinary (non-partial) derivative* is a full derivative, such as $dQ/dt$, where you take the derivative of all terms in $Q$ with respect to $t$. Here's an example of an ordinary differential equation, relating the charge $Q(t)$ in a circuit to the electromotive force $E(t)$ (that is, the voltage source connected to the circuit):

$$L\frac{d^2Q}{dt^2}+R\frac{dQ}{dt}+\frac{1}{C}Q=E(t)$$

Here, $Q$ is the charge, $L$ is the inductance of the circuit, $C$ is the capacitance of the circuit, and $E(t)$ is the electromotive force (voltage) applied to the circuit. This is an ordinary differential equation because only ordinary derivatives appear.

On the other hand, partial derivatives are taken with respect to only one variable, although the function depends on two or more. Here's an example of a partial differential equation (note the squiggly *d*'s):

$$\alpha^2\frac{\partial^2 u(x,t)}{\partial x^2}=\frac{\partial u(x,t)}{\partial t}$$

In this heat conduction equation, $\alpha$ is a physical constant of the system that you're trying to track the heat flow of, and $u(x, t)$ is the actual heat.

Note that $u(x, t)$ depends on both $x$ and $t$ and that both derivatives are partial derivatives — that is, the derivatives are taken with respect to one or the other of $x$ or $t$, but not both.

TIP

In this book, I focus on ordinary differential equations, because partial differential equations are usually the subject of more advanced texts. Never fear though: I promise to get you your fair share of partial differential equations.

## Classifying linear versus nonlinear equations

REMEMBER

Another way that you can classify differential equations is as *linear* or *nonlinear*. You call a differential equation *linear* if it exclusively involves linear terms (that is, terms to the power 1) of $y$, $y'$, $y''$, and beyond to $y^{(n)}$. For example, this equation is a linear differential equation:

$$L\frac{d^2Q}{dt^2}+R\frac{dQ}{dt}+\frac{1}{C}Q=E(t)$$

Note that this kind of differential equation usually will be written this way throughout this book. And this form makes the linear nature of this equation clear:

$$LQ'' + RQ' + \frac{1}{C}Q = E(t)$$

On the other hand, nonlinear differential equations involve nonlinear terms in any of $y$, $y'$, $y''$, up to $y^{(n)}$. The following equation, which describes the angle of a pendulum, is a nonlinear differential equation that involves the term $\sin\theta$ (not just $\theta$):

$$\frac{d^2\theta}{dt^2} + \frac{g}{L}\sin\theta = 0$$

Handling nonlinear differential equations is generally more difficult than handling linear equations. After all, it's often tough enough to solve linear differential equations without messing things up by adding higher powers and other nonlinear terms. For that reason, you'll often see scientists cheat when it comes to nonlinear equations. Usually they make an approximation that reduces the nonlinear equation to a linear one.

For example, when it comes to pendulums, you can say that for small angles, $\sin\theta \approx \theta$. This means that the following equation is the standard form of the pendulum equation that you'll find in physics textbooks:

$$\frac{d^2\theta}{dt^2} + \frac{g}{L}\theta = 0$$

As you can see, this equation is a linear differential equation, and as such, it's much more manageable. Yes, it's a cheat to use only small angles so that $\sin\theta \approx \theta$, but unless you cheat like that, you'll sometimes be reduced to using numerical calculations on a computer to solve nonlinear differential equations; obviously these calculations work, but it's much less satisfying than cracking the equation yourself (if you're a math geek like me).

# Solving First Order Differential Equations

Chapters 2, 3, and 4 take a look at differential equations of the form $f'(x) = f(x, y)$; these equations are known as first order differential equations because the derivative involved is of first order (for more on these types of equations, see the earlier section "Classifying equations by order."

First order differential equations are great because they're usually the most solvable. I show you all kinds of ways to handle first order differential equations in Chapters 2, 3, and 4. The following are some examples of what you can look forward to:

- As you know, first order differential equations look like this: $f'(x) = f(x, y)$. In the upcoming chapters, I show you how to deal with the case where $f(x, y)$ is linear in $x$ — for example, $f'(x) = 5x$ — and then nonlinear in $x$, as in $f'(x) = 5x^2$.
- You find out how to work with separable equations, where you can factor out all the terms having to do with $y$ on one side of the equation and all the terms having to do with $x$ on the other.
- I also help you solve first order differential equations in cool ways, such as by finding integrating factors to make more difficult problems simple.

REMEMBER

Direction fields, which I discuss earlier in this chapter, work only for equations of the type $f'(x) = f(x, y)$ — that is, where only the first derivative is involved — because the first derivative of $f(x)$ gives you the slope of $f(x)$ at any point (and, of course, connecting the slope line segments is what direction fields are all about).

# Tackling Second Order and Higher Order Differential Equations

As noted in the earlier section "Classifying equations by order," second order differential equations involve only the second derivative, $d^2y/dx^2$, also known as $y''$. In many physics situations, second order differential equations are where the action is.

For example, you can handle physics situations such as masses on springs or the electrical oscillations of inductor-capacitor circuits with a differential equation like this:

$$y'' - ay = 0$$

In Part II, I show you how to tackle second order differential equations with a large arsenal of tools, such as the *Wronskian matrix determinant,* which will tell you if there are solutions to a second (or higher) order differential equation. Other tools I introduce you to include the method of *undetermined coefficients* and the method of *variation of parameters.*

After first and second order differential equations, it's natural to want to keep the fun going, and that means you'll be dealing with higher order differential equations, which I also cover in Part II. With these high-end equations, you find terms like $d^ny/dx^n$, where $n > 2$.

TIP

The derivative $d^ny/dx^n$ is also written as $y^{(n)}$. Using the standard syntax, derivatives are written as $y'$, $y''$, $y'''$, $y^{iv}$, $y^{v}$, and so on. In general, the $n$th derivative of $y$ is written as $y^{(n)}$.

Higher order differential equations can be tough; many of them don't have solutions at all. But don't worry, because to help you solve them I bring to bear the wisdom of more than 300 years of mathematicians.

# Having Fun with Advanced Techniques

You discover dozens of tools in Part III of this book; all of these tools have been developed and proved powerful over the years. Laplace Transforms, Euler's method, integrating factors, numerical methods — they're all in this book.

These tools are what this book is all about — applying the knowledge of hundreds of years of solving differential equations. As you may know, differential equations can be broken down by type, and there's always a set of tools developed that allows you to work with whatever type of equation you come up with. In this book, you'll find a great many powerful tools that are just waiting to solve all of your differential equations — from the simplest to the seemingly impossible!

# Chapter 2

# Looking at Linear First Order Differential Equations

### *In This Chapter*

- Beginning with the basics of solving linear first order differential equations
- Using integrating factors
- Determining whether solutions exist for linear and nonlinear equations

As you find out in Chapter 1, a first order differential equation simply has a derivative of the first order. Here's what a typical first order differential equation looks like, where $f(t, y)$ is a function of the variables $t$ and $y$ (of course, you can use any variables here, such as $x$ and $y$ or $u$ and $v$, not just $t$ and $y$):

$$\frac{dy}{dt} = f(t,y)$$

In this chapter, you work with linear first order differential equations — that is, differential equations where the highest power of $y$ is 1 (you can find out the difference between linear and nonlinear equations in Chapter 1). For example:

$$\frac{dy}{dt} - 5$$

$$\frac{dt}{dt} = y + 1$$

$$\frac{dt}{dt} = 3y + 1$$

I provide some general information on nonlinear differential equations at the end of the chapter for comparison.

# First Things First: The Basics of Solving Linear First Order Differential Equations

In the following sections, I take a look at how to handle linear first order differential equations in general. Get ready to find out about initial conditions, solving equations that involve functions, and constants.

## Applying initial conditions from the start

REMEMBER

When you're given a differential equation of the form $dy/dt = f(t, y)$, your goal is to find a function, $y(t)$, that solves it. You may start by integrating the equation to come up with a solution that includes a constant, and then you apply an initial condition to customize the solution. Applying the initial condition allows you to select one solution among the infinite number that result from the integration. Sounds cool, doesn't it?

Take a look at this simple linear first order differential equation:

$$\frac{dy}{dt} = a$$

As you can see, $a$ is just a regular old number, meaning that this is a simple example to start with and to introduce the idea of initial conditions. How can you solve it? First of all, you may have noticed that another way of writing this equation is:

$$dy = a\,dt$$

This equation looks promising. Why? Well, because now you can integrate like this:

$$\int_{y_0}^{y} dy = \int_{x_0}^{t} a\ dt$$

Performing the integration gives you the following equation:

$$y - y_0 = at - at_0$$

You can combine $y_0 - at_0$ into a new constant, $c$, by adding $y_0$ to the right side of the equation, which gives you:

$$y = at + c$$

That was simple enough, right? And guess what? You're done! The solution to this differential equation is $y = at + c$.

So, for example, if $a = 3$ in the differential equation, here's the equation you would have:

$$\frac{dy}{dt} = 3$$

The solution for this equation is $y = 3t + c$.

TIP

Note that $c$, the result of integrating, can be any value, which leads to an infinite set of solutions: $y = 3t + 5$, $y = 3t + 6$, $y = 3t + 589{,}303{,}202$. How do you track down the value of $c$ that works for you? Well, it all depends on your initial conditions; for example, you may specify that the value of $y$ at $t = 0$ be 15. Setting this initial condition allows you to state the whole problem — differential equation and initial condition — as follows:

$$\frac{dy}{dt} = 3$$

$$y(0) = 15$$

Substituting the initial condition, $y(0) = 15$, into the solution $y = 3t + c$ gives you the following equation:

$$y(t) = 3t + 15$$

## Stepping up to solving differential equations involving functions

Of course, $dy/dt = 3$ (the example from the previous section) isn't the most exciting differential equation. However, it does show you how to solve a differential equation using integration and how to apply an initial condition. The next step is to solve linear differential equations that involve functions of $t$ rather than just a simple number.

REMEMBER

This type of differential equation still contains only $dy/dt$ and terms of $t$, making it easy to integrate. Here's the basic form:

$$\frac{dy}{dt} = g(t)$$

where $g(t)$ is some function of $t$.

Here's an example of this type of differential equation:

$$\frac{dy}{dt} = t^3 - 3t^2 + t$$

Well, heck, that's easy too; you simply rearrange to get this:

$$dy = t^3\,dt - 3t^2\,dt + t\,dt$$

Then you can integrate to get this equation:

$$y = \frac{t^4}{4} - t^3 + \frac{t^2}{2} + c$$

## Adding a couple of constants to the mix

The next step up from equations such as $dy/dx = a$ or $dy/dt = g(t)$ are equations of the following form, which involve $y$, $dy/dt$, and the constants $a$ and $b$:

$$\frac{dy}{dt} = ay - b$$

How do you handle this equation and find a solution? Using some handy algebra, you can rewrite the equation like this:

$$\frac{dy/dt}{y - (b/a)} = a$$

Integrating both sides gives you the following equation:

$$\ln |\, y - (b/a)\, | = at + c$$

where $c$ is an arbitrary constant. Now get $y$ out of the natural logarithm, which gives you:

$$y = (b/a) + de^{at}$$

where $d = e^c$. And that's it! You're done. Good job!

# Solving Linear First Order Differential Equations with Integrating Factors

REMEMBER

Sometimes integrating linear first order differential equations isn't as easy as it is in the examples earlier in this chapter. But it turns out that you can often convert general equations into something that's easy to integrate if you find an *integrating factor,* which is a function, $\mu(t)$. The idea here is to multiply the differential equation by an integrating factor so that the resulting equation can easily be integrated and solved.

In the following sections, I provide tips and tricks for solving for an integrating factor and plugging it back into different types of linear first order equations.

## Solving for an integrating factor

In general, first order differential equations don't lend themselves to easy integration, which is where integrating factors come in. How does the method of integrating factors work? To understand, say, for example, that you have this linear differential equation:

$$\frac{dy}{dt} + 2y = 4$$

First, you multiply the previous equation by $\mu(t)$, which is a stand-in for the undetermined integrating factor, giving you:

$$\mu(t)\frac{dy}{dt} + 2\mu(t)\,y = 4\mu(t)$$

Now you have to choose $\mu(t)$ so that you can recognize the left side of this equation as the derivative of some expression. This way it can easily be integrated.

Here's the key: The left side of the previous equation looks very much like differentiating the product $\mu(t)y$. So try to choose $\mu(t)$ so that the left side of the equation is indeed the derivative of $\mu(t)y$. Doing so makes the integration easy.

The derivative of $\mu(t)y$ by $t$ is:

$$\frac{d\left[\mu(t)\,y\right]}{dt} = \mu(t)\frac{dy}{dt} + \frac{d\mu(t)\,y}{dt}$$

Comparing the previous two equations term by term gives you:

$$\frac{d\mu(t)}{dt} = 2\mu(t)$$

Hey, not bad. Now you're making progress! This is a differential equation you can solve. Rearranging the equation so that all occurrences of $\mu(t)$ are on the same side gives you:

$$\frac{d\mu(t)/dt}{\mu(t)} = 2$$

Now the equation can be rearranged to look like this:

$$\frac{d\mu(t)}{\mu(t)} = 2\,dt$$

Fine work. Integration gives you:

$$\ln|\mu(t)| = 2t + b$$

where $b$ is an arbitrary constant of integration.

Now it's time for some exponentiating. Exponentiating both sides of the equation gives you:

$$\mu(t) = ce^{2t}$$

where $c$ is an arbitrary constant.

So that's it — you've solved for the integrating factor! It's $\mu(t) = ce^{2t}$.

## Using an integrating factor to solve a differential equation

After you solve for an integrating factor, you can plug that factor into the original linear differential equation as multiplied by $\mu(t)$. For instance, take your original equation from the previous section:

$$\mu(t)\frac{dv}{dt} + 2\mu(t)\,y = 4\mu(t)$$

and plug in the integrating factor to get this equation:

$$ce^{2t}\frac{dy}{dt} + 2\,ce^{2t}\,y = 4\,ce^{2t}$$

TIP

Note that $c$ drops out of this equation when you divide by $c$, so you get the following equation (because you're just looking for an arbitrary integrating factor, you could also set $c = 1$):

$$e^{2t}\frac{dy}{dt} + 2\,e^{2t}\,y = 4\,e^{2t}$$

REMEMBER

When you use an integrating factor, you attempt to find a function $\mu(t)$ that, when multiplied on both sides of a differential equation, makes the left side into the derivative of a product. Figuring out the product allows you to solve the differential equation.

In the previous example, you can now recognize the left side as the derivative of $e^{2t}\,y$. (If you can't recognize the left side as a derivative of some product, in general, it's time to go on to other methods of solving the differential equation).

In other words, the differential equation has been conquered, because now you have it in this form:

$$\frac{d\left(e^{2t}\,y\right)}{dt} = 4e^{2t}$$

You can integrate both sides of the equation to get this:

$$e^{2t}y = 2e^{2t} + c$$

And, finally, you can solve for $y$ with your handy algebra skills:

$$y = 2 + ce^{-2t}$$

You've got yourself a solution. Beautiful.

REMEMBER

The use of an integrating factor isn't always going to help you; sometimes, when you use an integrating factor in a linear differential equation, the left side isn't going to be recognizable as the derivative of a product of functions. In that case, where integrating factors don't seem to help, you have to turn to other methods. One of those methods is to determine whether the differential equation is *separable,* which I discuss in Chapter 3.

## Moving on up: Using integrating factors in differential equations with functions

Now you're going to take integrating factors to a new level. Check out this linear equation, where $g(t)$ is a function of $t$:

$$\frac{dy}{dt} + ay = g(t)$$

This one's a little more tricky. However, using the same integrating factor from the previous two sections, $e^{at}$ (remember that the $c$ dropped out), works here as well. After you multiply both sides by $e^{at}$, you get this equation:

$$e^{at}\frac{dy}{dt} + a\, e^{at} y = e^{at} g(t)$$

Now you can recast this equation in the following form:

$$\frac{d\left(e^{at} y\right)}{dt} = e^{at} g(t)$$

To integrate the function $g$, I use $s$ as the variable of integration. Integration gives you this equation:

$$e^{at} y = \int e^{as} g(s)\, ds + c$$

You can solve for $y$ here, which gives you the following equation:

$$y = e^{-at} \int e^{as} g(s)\, ds + ce^{-at}$$

And that's it! You've got your answer!

WARNING!

Of course, solving this equation depends on whether you can calculate the integral in the previous equation. If you can do it, you've solved the differential equation. Otherwise, you may have to leave the solution in the integral form.

## Trying a special shortcut

In this section, I give you a shortcut for solving some particular differential equations. Ready? Here's the tip: In general, the integrating factor for an equation in this form:

$$\frac{dy}{dt} + ay = g(t)$$

is this:

$$\mu(t) = \exp \int a \, dt$$

In this equation, exp($x$) means $e^x$.

As an example, try solving the following differential equation with the shortcut:

$$\frac{dy}{dt} + \frac{1}{2} y = 4 + t$$

Assume that the initial condition is

$$y = 8, \text{ when } t = 0$$

This equation is an example of the general equation solved in the previous section. In this case, $g(t) = 4 + t$, and $a = ½$.

Using $a$, you find that the integrating factor is $e^{t/2}$, so multiply both sides of equation by that factor:

$$e^{t/2}\frac{dy}{dt} + \frac{e^{t/2}}{2} y = 4e^{t/2} + te^{t/2}$$

Now you can combine the two terms on the left to give you this equation:

$$\frac{d\left(e^{t/2} y\right)}{dt} = 4e^{t/2} + te^{t/2}$$

All you have to do now is integrate this result. The term on the left and the first term on the right are no problem. The last term on the right is another story.

TIP

You can use *integration by parts* to integrate this term. Integration by parts works like this:

$$\int_a^b f(x) g'(x) \, dx = f(b) g(b)$$

$$- f(a) g(a) - \int_a^b f'(x) g(x) \, dx$$

Applying integration by parts to the last term on the right, and integrating the others, gives you:

$$e^{t/2} y = 8e^{t/2} + 2\,t\,e^{t/2} - 4e^{t/2} + c$$

where $c$ is an arbitrary constant, set by the initial conditions. Dividing by $e^{at}$ gives you this equation:

$$y = 4 + 2t + ce^{-at}$$

By applying the initial condition, $y(0) = 8$, you get

$$y(0) = 8$$
$$8 = 4 + c$$

Or $c = 4$. So the general solution of the differential equation is:

$$y = 4 + 2t + 4e^{-t/2}$$

TIP

In Chapter 1, I explain that direction fields are great tools for visualizing differential equations. You can see a direction field for the previously noted general solution in Figure 2-1.

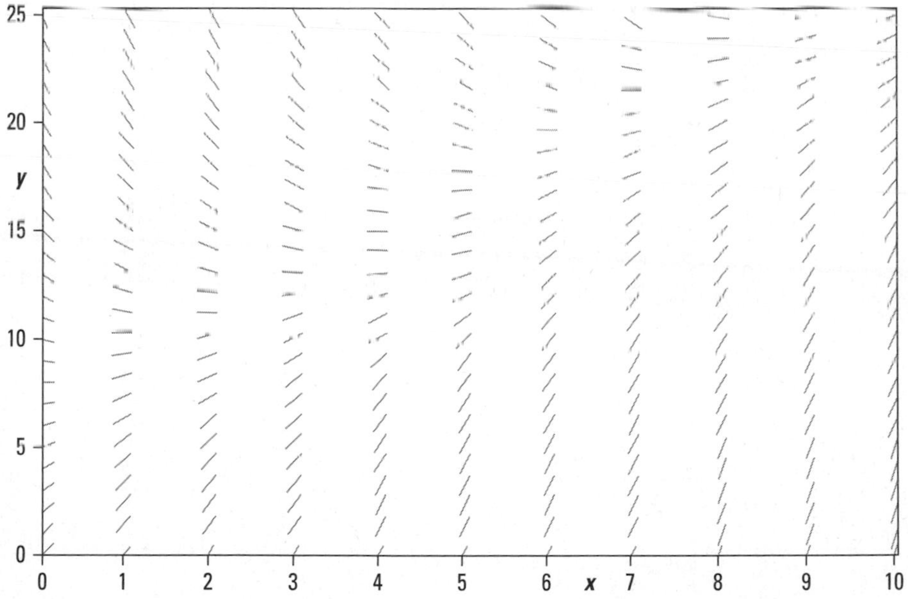


**Figure 2-1:** The direction field of the general solution.

Connecting the slanting lines in a direction field gives you a graph of the solution. You can see a graph of this solution in Figure 2-2.

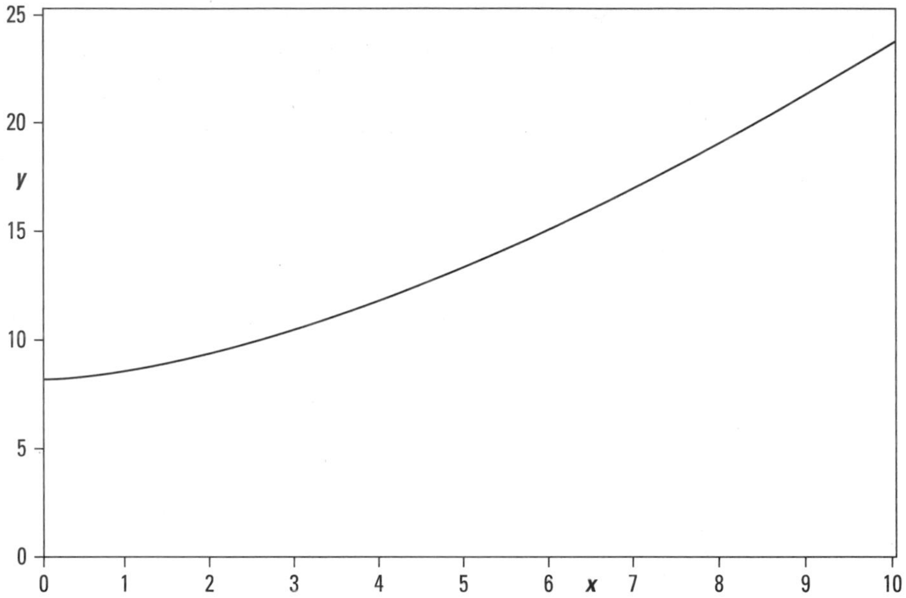


**Figure 2-2:** The graph of the general solution.

## Solving an advanced example

I think you're ready for another, somewhat more advanced, example. Try solving this differential equation to show that you can have different integrating factors:

$$t\frac{dy}{dt} + 2y = 4t^2$$

where $y(1) = 4$.

To solve, first you have to find an integrating factor for the equation. To get it into the form:

$$\frac{dy}{dt} + ay = g(t)$$

you have to divide both sides by $t$, which gives you this equation:

$$\frac{dy}{dt} + \frac{2}{t}y = 4t$$

To find the integrating factor, use the shortcut equation from the previous section, like this:

$$\mu(t) = \exp \int a \, dt = \exp \int \frac{2}{t} \, dt$$

Performing the integral gives you this equation:

$$\mu(t) = \exp \int \frac{2}{t} \, dt = e^{2\ln|t|} = t^2$$

So the integrating factor here is $t^2$, which is a new one. Multiplying both sides of the equation by the integrating factor, $\mu(t) = t^2$, gives you:

$$t^2 \frac{dy}{dt} + 2ty = 4t^3$$

Because the left side is a readily apparent derivative, you can also write it in this form:

$$\frac{d(yt^2)}{dt} = 4t^3$$

Now simply integrate both sides to get:

$$yt^2 = t^4 + c$$

Finally you get:

$$y = t^2 + \frac{c}{t^2}$$

where $c$ is an arbitrary constant of integration.

Now you can plug in the initial condition $y(1) = 4$, which allows you to see that $c = 3$. And that helps you come to this solution:

$$y = t^2 + \frac{3}{t^2}$$

And there you have it. You can see a direction field for the many general solutions to this differential equation in Figure 2-3.

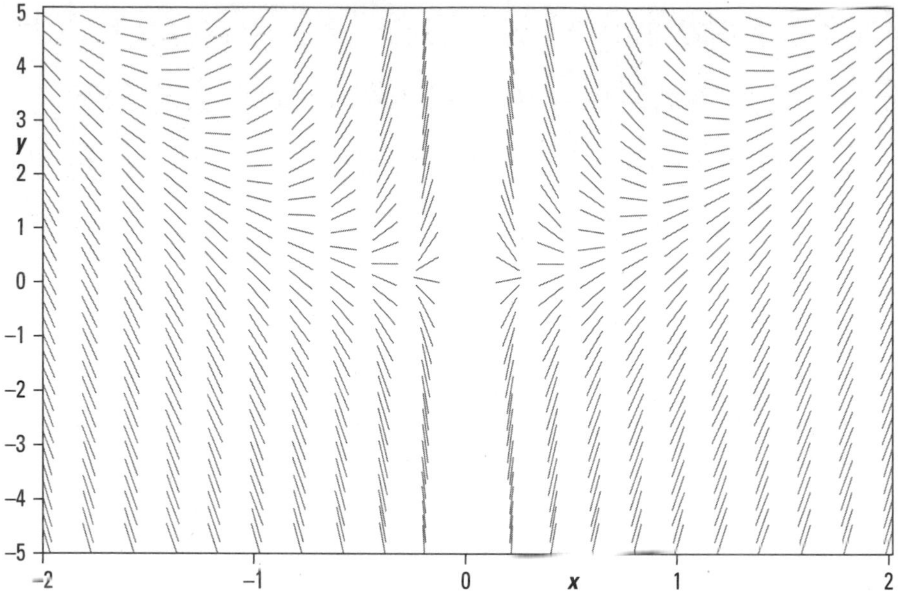


**Figure 2-3:** The direction field of a more advanced solution.

You can see this function graphed in Figure 2-4.

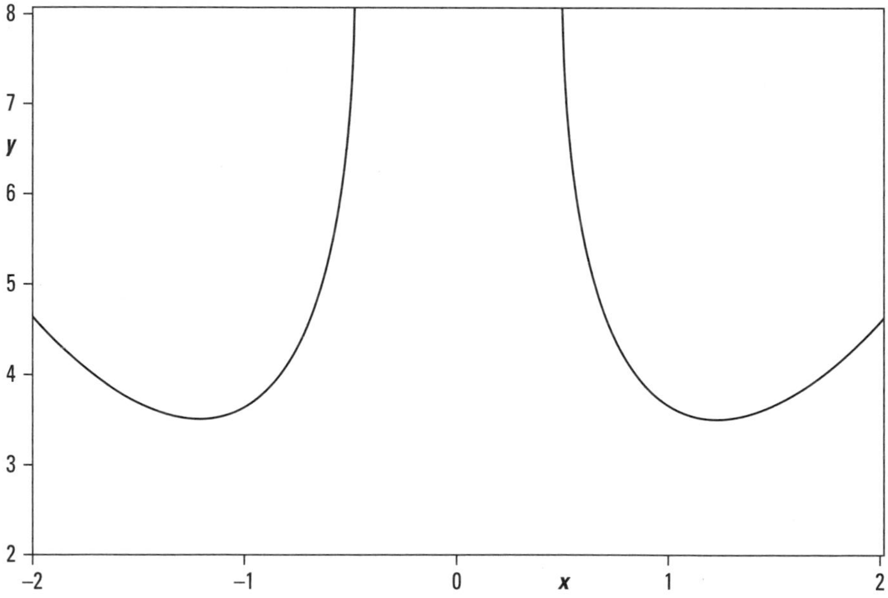


**Figure 2-4:** The graph of a more advanced solution.

# Determining Whether a Solution for a Linear First Order Equation Exists

I show you how to deal with different kinds of linear first order differential equations earlier in this chapter, but the fact remains that not all linear differential equations actually do have a solution.

Luckily, a theorem exists that tells you when a given linear differential equation with an initial condition has a solution. That theorem is called the *existence and uniqueness theorem.*

This theorem is worth knowing. After all, if a differential equation doesn't have a solution, what use is it to search for a solution? In other words, this theorem represents another way to tackle linear first order differential equations.

## Spelling out the existence and uniqueness theorem for linear differential equations

In this section, I explain what the existence and uniqueness theorem for linear differential equations says. Before I continue, however, note that a *continuous function* is a function for which small changes in the input result in small changes in the output (for example, $f(x) = 1/x$ is not continuous at $x = 0$). Without further ado, here's the existence and uniqueness theorem:

**If there is an interval $I$ that contains the point $t_o$, and if the functions $p(x)$ and $g(x)$ are continuous on that interval, and if you have this differential equation:**

$$\frac{dy}{dx}+p(x)y=g(x)$$

**then there exists a unique function, $y(x)$, that is the solution to that differential equation for each $x$ in interval $I$ that also satisfies this initial condition:**

$$y(t_o) = y_o$$

**where $y_o$ is an arbitrary initial value.**

In other words, this theorem says that a solution *exists* and that the solution is *unique.*

## Finding the general solution

Thinking about the theorem in the previous section begs the question: What is the general solution to the following linear differential equation?

$$\frac{dy}{dx} + p(x)\,y = g(x)$$

Note that this differential equation has a function $p(x)$ and $g(x)$, which provides a more complex situation. So you can't use the simple form I explain in the earlier section "Adding a couple of constants to the mix," where $a$ and $b$ are constants like this:

$$\frac{dy}{dx} = ay - b$$

The solution here is:

$$y = (b/a) + ce^{at}$$

Now you face a more complex situation, with functions $p(x)$ and $g(x)$. A general solution to the general equation does exist, and here it is:

$$y = \frac{\int \mu(s)\,g(s)\,ds + c}{\mu(t)}$$

where the integrating factor is the following:

$$\mu(t) = \exp \int p(t)\,dt$$

The integrals in these equations may not be possible to perform, of course. But together, the equations represent the general solution.

Note that for linear differential equations, the solution, if there is one, is completely specified, up to a constant of integration, as in the solution you get in the earlier section "Solving an advanced example":

$$y = t^2 + \frac{c}{t^2}$$

where $c$ is a constant of integration.

You can't necessarily say the same thing about nonlinear differential equations — they may have solutions of completely different forms, not just differing in the value of a constant. Because the solution to a linear differential equation has one form, differing only by the value of a constant, those solutions are referred to as *general solutions.* This term isn't used when discussing nonlinear differential equations, which may have multiple solutions of completely different forms. I discuss nonlinear first order differential equations later in this chapter.

## Checking out some existence and uniqueness examples

In this section, I include a few examples to help you understand the existence and uniqueness theorem for linear differential equations.

### Example 1

Apply the existence and uniqueness theorem to the following equation to show that there exists a unique solution:

$$\frac{dy}{dx} = \frac{4x}{(y-5)}$$

Just kidding! This equation isn't linear because the term $(y - 5)$ is in the denominator of the right side. And, of course, because the equation isn't linear, the existence and uniqueness theorem doesn't apply. Did you catch that?

### Example 2

Try this differential equation (which I promise is linear!). Does a unique solution exist?

$$\frac{dy}{dx} + 2y = 4x^2$$

where $y(1) = 2$.

The equation is already in the correct form:

$$\frac{dy}{dx} + p(x)\,y = g(x)$$

where $p(x) = 2$ and $g(x) = 4x^2$.

Note that $p(x)$ and $g(x)$ are continuous everywhere, so there's a general solution that's valid on the interval, $-\infty < x < \infty$.

In particular, the initial condition is $y(1) = 2$, which is definitely inside the interval that $p(x)$ and $g(x)$ are continuous (everything is inside that interval). So, yes, there exists a solution to the initial value problem.

### Example 3

Now take a look at this equation, which is similar to the example in the previous section, and determine whether a unique solution exists:

$$x\frac{dy}{dx} + 2y = 4x^2$$

where $y(1) = 2$.

The next step is to put the equation into this form:

$$\frac{dy}{dx} + p(x)\,y = g(x)$$

Here's what the equation should look like:

$$\frac{dy}{dx} + \frac{2y}{x} = 4x$$

In other words:

$$p(x) = \frac{2}{x}$$

and

$$g(x) = 4x$$

Note that $p(x)$ and $g(x)$ aren't continuous everywhere. In particular, $p(x)$ is discontinuous at $x = 0$, which makes the interval in which $p(x)$ and $g(x)$ are continuous on the interval $0 > x$ and $0 < x$.

Because the initial condition here is $y(1) = 2$, the point of interest is $x = 1$, which is inside the interval where $p(x)$ and $g(x)$ are continuous. Therefore, by the existence and uniqueness theorem, the initial value problem indeed has a unique solution. Cool, huh?

# Figuring Out Whether a Solution for a Nonlinear Differential Equation Exists

In the previous sections of this chapter, I cover linear first order differential equations in detail. But you may be wondering: Is there such a thing as a nonlinear differential equation? You bet there is! A *nonlinear differential equation* simply includes nonlinear terms in y, y′, y″, and so on. Nonlinear equations are pretty tough, so I don't delve into them a lot in this book. But I do want to discuss one important theorem related to solving these equations.

You see, the existence and uniqueness theorem (which you use for linear equations, and which I cover earlier in this chapter) is analogous to another theorem that's used for nonlinear equations. I explain this theorem and show some examples in the following sections.

## The existence and uniqueness theorem for nonlinear differential equations

Here's the existence and uniqueness of solutions for nonlinear equations:

**Say that you have a rectangle $R$ that contains the point $(t_o, y_o)$ and that the functions $f$ and $df/dy$ are continuous in that rectangle. Then, in an interval $t_o - h < t < t_o + h$ contained in $R$, there's a unique solution to the initial value problem:**

$$\frac{dy}{dt} = f(t,y)\,,\ y(t_0) = y_0$$

Note that this theorem discusses the continuity of both $f$ and $df/dy$ instead of the continuity of both $p(x)$ and $g(x)$. Like the first theorem in this chapter, this theorem guarantees the existence of a unique solution if its conditions are met.

Here's another note: If the differential equation in question actually is linear, the theorem reduces to the first theorem in this chapter. In that case, $f(t, y) = -p(t)y + g(t)$ and $df/dy = -p(t)$. So demanding that $f$ and $df/dy$ be continuous is the same as saying that $p(t)$ and $g(t)$ be continuous.

Here's a side note that many differential equations books won't tell you: The first theorem in this chapter guarantees a unique solution, but it's actually a little tighter than it needs to be in order to guarantee just a solution (which isn't necessarily unique). In fact, you can show that there's a solution — but not that it's unique — to the nonlinear differential equation merely by proving that $f$ is continuous.

## A couple of nonlinear existence and uniqueness examples

In the following sections, I provide two examples that put the nonlinear existence and uniqueness theorem into action.

### Example 1

Determine what the two theorems in this chapter have to say about the following differential equation as far as its solutions go:

$$\frac{dy}{dx} = \frac{5x^2 + 9x + 6}{2(y-4)}$$

where $y(0) = -1$.

Well, as you can see, this is a nonlinear equation in *y*. So the first theorem, which deals only with linear differential equations, has nothing to say about it.

That means you need the nonlinear theorem. Note that for this theorem:

$$f(x,y) = \frac{5x^2 + 9x + 6}{2(y-4)}$$

and

$$\frac{df}{dy} = -\frac{5x^2 + 9x + 6}{2(y-4)^2}$$

These two functions, *f* and *df*/*dy*, are continuous, except at $y = 4$.

So you can draw a rectangle around the initial condition point, (0, –1) in which both *f* and *df*/*dy* are continuous. And the existence and uniqueness theorem for nonlinear equations guarantees that this differential equation has a solution in that rectangle.

### Example 2

Now determine what the existence and uniqueness theorems say about this differential equation:

$$\frac{dy}{dx} = y^{1/5}$$

where $y(1) = 0$.

Clearly, this equation isn't linear, so the first theorem is no good. Instead you have to try the second theorem. Here, *f* is:

$$f(x, y) = y^{1/5}$$

and *df*/*dy* is:

$$\frac{df}{dy} = \frac{y^{-4/5}}{5}$$

Now you know that $f(x, y)$ is continuous at the initial condition point given by:

$$y(1) = 0$$

But *df*/*dy* isn't continuous at this point. The upshot is that neither the first theorem nor the second theorem have anything to say about this initial value problem. On the other hand, a solution to this differential equation is still guaranteed because $f(x, y)$ is continuous. However, it doesn't guarantee the uniqueness of that solution.

# Chapter 3

# Sorting Out Separable First Order Differential Equations

### In This Chapter

- Figuring out the fundamentals of separable differential equations
- Applying separable differential equations to real life
- Advancing with partial fractions

Some rocket scientists call you, the Consulting Differential Equation Expert, into their headquarters.

"We've got a problem," they explain. "Our rockets are wobbling because we can't solve their differential equation. All the rockets we launch wobble and then crash!"

They show you to a blackboard with the following differential equation:

$$\frac{dy}{dx} = \frac{x^2}{2 - y^2}$$

"It's not linear," the scientists cry. "There's a $y^2$ in there!"

"I can see that," you say. "Fortunately, it is separable."

"Separable? What does that mean?" they ask.

"Separable means that you can recast the equation like this, where $x$ is on one side and $y$ is on the other," you say while showing them the following equation on your clipboard:

$$(2 - y^2)\, dy = x^2\, dx$$

"You can integrate the equation with respect to $y$ on one side, and $x$ on the other," you say.

"We never thought of that. That was too easy."

That's what this chapter covers: separable first order differential equations. (First order equations, as I note in Chapter 1, have derivatives that go up only to the first order.) I explain the basics of separable equations here, such as determining the difference between linear and nonlinear separable equations and figuring out different types of solutions, such as implicit and explicit. I also introduce you to a fancy method for solving separable equations involving partial fractions. Finally, I show you a couple of real world applications for separable equations. When you're an expert at these equations, you too can solve problems for rocket scientists.

# Beginning with the Basics of Separable Differential Equations

REMEMBER

*Separable differential equations,* unlike general linear equations in Chapter 2, let you separate variables so only variables of one kind appear on one side, and only variables of another kind appear on the other. Say, for example, that you have a differential equation of the following form, in which $M$ and $N$ are functions:

$$M(x,y)+N(x,y)\frac{dy}{dx}=0$$

And furthermore, imagine that you could reduce this equation to the following form, where the function $M$ depends only on $x$ and the function $N$ depends only on $y$:

$$M(x)+N(y)\frac{dy}{x}=0$$

This equation is a separable equation; in other words, you can separate the parts so that only $x$ appears on one side, and only $y$ appears on the other. You write the previous equation like this:

$$M(x)\,dx+N(y)\,dy=0$$

Or in other words:

$$M(x)\,dx=-N(y)\,dy$$

If you can separate a differential equation, all that's left to do at that point is to integrate each side (assuming that's possible). Note that the general form of a separable differential equation looks like this:

$$M(x)+N(y)\frac{dy}{dx}=0$$

However, nothing here says that $N(y)$ has to be linear in $y$. For example, consider this separable differential equation that isn't linear:

$$x + y^2 \frac{dy}{dx} = 0$$

And if you're still not convinced, check out this one, which is also separable but not linear:

$$x^9 + \left(1 - y^3\right)\frac{dy}{dx} = 0$$

In the following sections, I ease you into linear separable equations before tackling nonlinear separable equations. I also show you a trick for turning nonlinear equations into linear equations. (It's so cool that it'll impress all your friends!)

## Starting easy: Linear separable equations

To get yourself started with linear separable equations, say that you have this differential equation:

$$\frac{dy}{dx} - x^2 = 0$$

This equation qualifies as linear. This also is an easily separated differential equation. All you have to do is put it into this form:

$$dy = x^2\,dx$$

And now you should be able to see the idea behind solving separable differential equations immediately. You just have to integrate, which gives you this equation:

$$y = \frac{x^3}{3} + c$$

where $c$ is an arbitrary constant. There's your solution! How easy was that?

## Introducing implicit solutions

Not all separable equation solutions are going to be as easy as the one in the previous section. Sometimes finding a solution in the $y = f(x)$ format isn't terribly easy to get. Mathematicians refer to a solution that isn't in the form $y = f(x)$ as an *implicit solution.* Coming up with such a solution is often the best you can do, because solving a separable differential equation involves integrating both sides of the equation, and there's no guarantee that the integration will come out cleanly. (The form $y = f(x)$ is known as an *explicit solution;* I show you how to find an explicit solution from an implicit solution in the next section.)

Try this differential equation to see what I mean:

$$\frac{dy}{dx} = \frac{x^2}{2 - y^2}$$

How about it? One of the first things that should occur to you is that this isn't a linear differential equation, so the techniques in the first part of this chapter won't help. However, you'll probably notice that you can write this equation as:

$$(2 - y^2)\, dy = x^2\, dx$$

As you can see, this is a separable differential equation because you can put $y$ on one side and $x$ on the other. You can also write the differential equation like this:

$$-x^2 + (2 - y^2)\frac{dy}{dx} = 0$$

You can cast this particular equation in terms of a derivative of $x$, and then you integrate with respect to $x$ to solve it. After integration, you wind up with the following:

$$-x^2 = \frac{d(-x^3/3)}{dx}$$

REMEMBER

Note that:

$$(2 - y^2)\frac{dy}{dx} = \frac{d(2y - y^3/3)}{dx}$$

because of the *chain rule,* which says that:

$$\frac{df}{dx} = \frac{df}{dy}\frac{dy}{dx}$$

So now you can write the original equation like this:

$$-x^2 + (2 - y^2)\frac{dy}{dx} = \frac{d}{dx}\left(\frac{-x^3}{3} + 2y - \frac{y^3}{3}\right) = 0$$

TIP

If the derivative of the term on the right is 0, it must be a constant this way:

$$\left(\frac{-x^3}{3} + 2y - \frac{y^3}{3}\right) = c$$

Finally, multiplying by 3 gives you the following implicit solution to your original separable equation:

$$-x^3 + 6y - y^3 = c$$

To see how the solutions look graphically, check out the direction field for this differential equation in Figure 3-1. (I introduce direction fields in Chapter 1.)

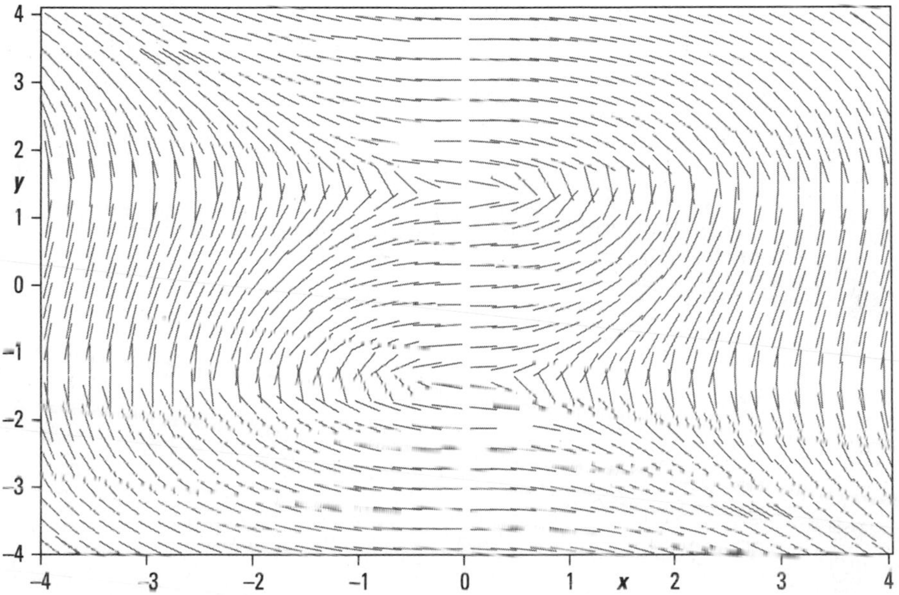


**Figure 3-1:** The direction field of a nonlinear separable equation.

## Finding explicit solutions from implicit solutions

The implicit solution in the previous section, with terms in $y$ and $y^3$, isn't terribly easy to cram into the $y = f(x)$ format. In this section, you discover that you can find an explicit solution to a separable equation by using a *quadratic equation,* which is the general solution to polynomials of order two.

Try another, somewhat more tractable problem. Solve this differential equation:

$$\frac{dy}{dx} = \frac{9x^2 + 6x + 4}{2(y-1)}$$

where $y(0) = -1$.

If this differential equation were of the following form:

$$\frac{dy}{dx} = 9x^2 + 6x + 4$$

there would be no problem. After all, you would just integrate. But you've probably noticed that pesky $2(y-1)$ term in the denominator on the right side. Fortunately, you may also realize that this is a separable differential equation because you can put $y$ on one side and $x$ on the other. Simply write the equation like this:

$$2(y-1)\,dy = (9x^2 + 6x + 4)\,dx$$

Now you integrate to get this equation:

$$y^2 - 2y = 3x^3 + 3x^2 + 4x + c$$

Using the initial condition, $y(0) = -1$, substitute $x = 0$ and $y = -1$ to get the following:

$$1 + 2 = c$$

Now you can see that $c = 3$ and that the implicit solution to the separable equation is:

$$y^2 - 2y = 3x^3 + 3x^2 + 4x + 3$$

If you want to find the explicit solution to this and similar separable equations, simply solve for $y$ with the quadratic equation because the highest power of $y$ is 2. Solving for $y$ using the quadratic formula gives you:

$$y = 1 \pm \sqrt{3x^3 + 3x^2 + 4x + 4}$$

You have two solutions here: one where the addition sign is used and one where the subtraction sign is used. To match the initial condition that $y(0) = -1$, however, only one solution will work. Which one? The one using the subtraction sign:

$$y = 1 - \sqrt{3x^3 + 3x^2 + 4x + 4}$$

In this case, the solution with the subtraction is valid as long as the expression under the square root is positive — in other words, as long as $x > -1$.

You can see the direction field for the general solutions to this differential equation in Figure 3-2.

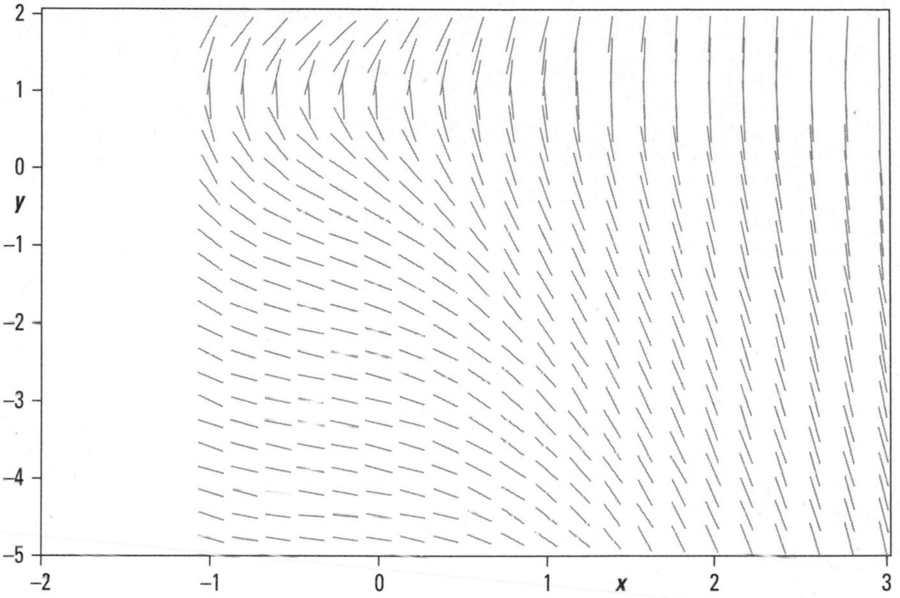


**Figure 3-2:** The direction field of a separable equation with initial conditions.

As I note in Chapter 1, connecting the slanting lines in a direction field gives you a graph of the solution. You can see a graph of this particular function in Figure 3-3.

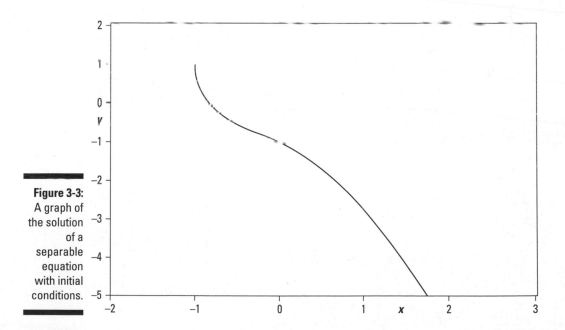

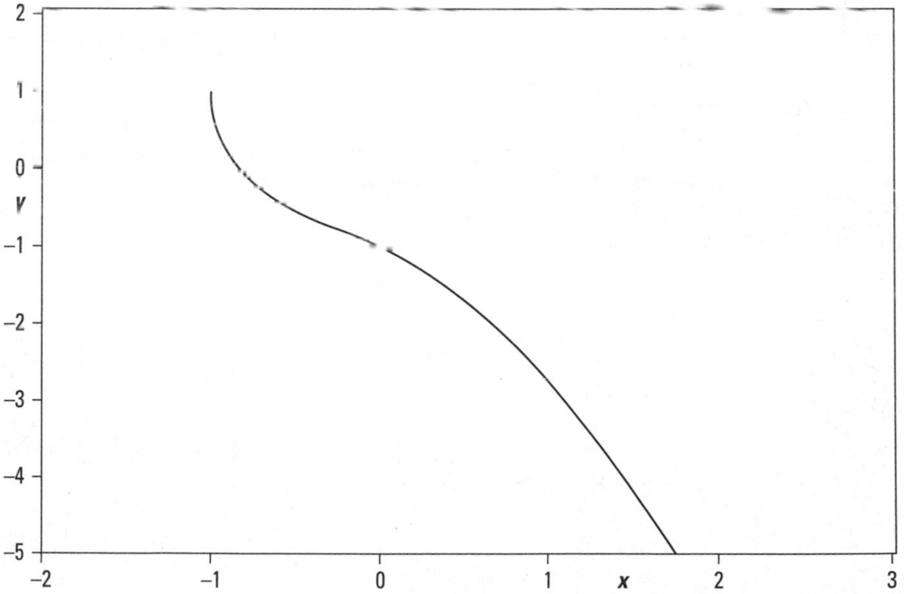


**Figure 3-3:** A graph of the solution of a separable equation with initial conditions.

## Tough to crack: When you can't find an explicit solution

Most of the time, you can find an explicit solution from an implicit solution. But every once in a while, getting an explicit solution is pretty tough to do. Here's an example:

$$\frac{dy}{dx} = \frac{y\sin x}{(1 + 2y^2)}$$

where $y(0) = 1$.

As you get down to work (bringing to bear all your differential equation skills!), the first thing that may strike you is that this equation isn't linear. But, you'll also likely note that it's separable. So simply separate the equation into $y$ on the left and $x$ on the right, which gives you this equation:

$$\frac{(1 + 2y^2)\,dy}{y} = \sin x \; dx$$

This equation subsequently becomes

$$\frac{dy}{y} + 2y \; dy = \sin x \; dx$$

Now you can integrate to get this:

$$\ln|y| + y^2 = -\cos x + c$$

Next, take a look at the initial condition: $y(0) = 1$. Plugging that condition into your solution gives you this equation:

$$0 + 1 = -1 + c$$

or

$$c = 2$$

So your solution to the initial separable equation is:

$$\ln|y| + y^2 = -\cos x + 2$$

REMEMBER

This is an implicit solution, not an explicit solution, which would be in terms of $y = f(x)$. In fact, as you can see from the form of this implicit solution, getting an explicit solution would be no easy task.

TIP

However, never fear the implicit solution! You still can use numerical or graphical methods to deal with such solutions. For instance, take a look at the direction field for this differential equation, which indicates what the integral curves look like, in Figure 3-4.

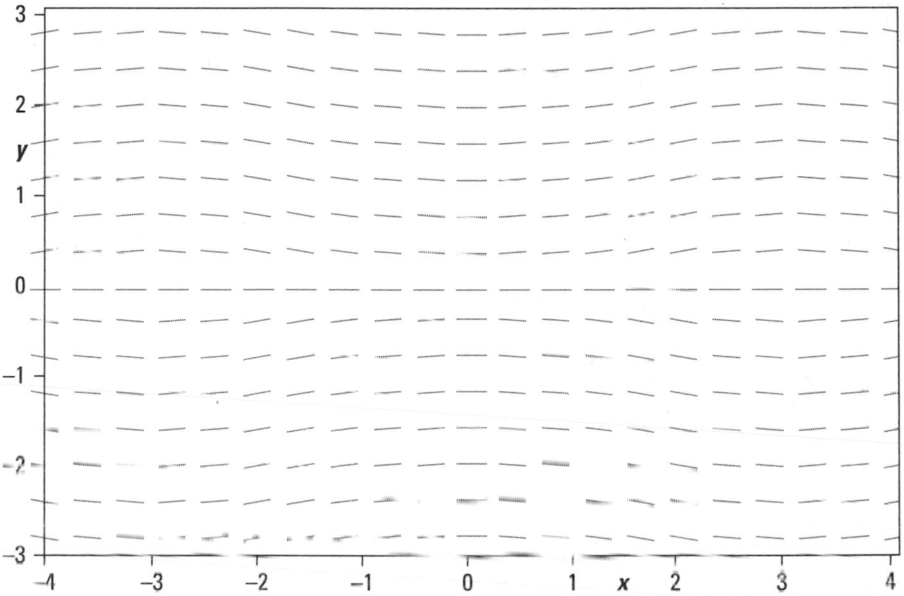


**Figure 3-4:** The direction field of a separable equation with a hard-to-find explicit solution.

## A neat trick: Turning nonlinear separable equations into linear separable equations

TIP

In this section, I introduce you to a neat trick that helps with some differential equations. With it, you can make a linear equation out of a seemingly nonlinear one. All you have to do to use this trick is to substitute the following equation, in which $v$ is a variable:

$$y = xv$$

In some cases, the result is a separable equation.

As an example, try solving this differential equation:

$$\frac{dy}{dx} = \frac{2y^4 + x^4}{xy^3}$$

At first glance, this equation doesn't look separable. In fact, even if you break it out into two fractions, it still doesn't look separable:

$$\frac{dy}{dx}=\frac{2y}{x}+\frac{x^3}{y^3}$$

What do you do now? Keep reading to find out.

### Knowing when to substitute

REMEMBER

You can use the trick of setting $y = xv$ when you have a differential equation that's of the following form:

$$\frac{dy}{dx}=f(x,y)$$

when $f(x, y) = f(tx, ty)$, where $t$ is a constant.

You can see that substitution is possible, because substituting $tx$ and $ty$ into this differential equation gives you the following result:

$$\frac{dy}{dx}=\frac{2t^4y^4+t^4x^4}{txt^3y^3}$$

which breaks down to:

$$\frac{dy}{dx}=\frac{2y^4+x^4}{xy^3}$$

Substituting $y = xv$ into this differential equation gives you:

$$v+x\frac{dv}{dx}=\frac{2(xv)^4+x^4}{x(xv)^3}$$

This equation now can be simplified to look like this:

$$x\frac{dv}{dx}=\frac{v^4+1}{v^3}$$

You now have a separable equation!

### Separating and integrating

Continuing with the example from the previous section, you can now separate the terms, which gives you:

$$\frac{dx}{x}=\frac{v^3\,dv}{v^4+1}$$

After you integrate both sides, you get the following equation:

$$\ln(x) = \frac{\ln(v^4 + 1)}{4} + c$$

where $c$ is a constant of integration. Bearing in mind that, where $k$ is a constant:

$$\ln(x) + \ln(k) = \ln(kx)$$

and that:

$$n \ln(x) = \ln(x^n)$$

you get:

$$v^4 + 1 = (kx)^4$$

where:

$$c = -\ln(k)$$

Where does all this get you? You're ready to substitute with the following:

$$v = \frac{y}{x}$$

This substitution gives you:

$$\left(\frac{y}{x}\right)^4 + 1 = (kx)^4$$

So:

$$y^4 + x^4 = mx^8$$

where $m = k^4$. And solving for $y$ gives you the following:

$$y = (mx^8 - x^4)^{1/4}$$

And there's your solution. Nice work!

# Trying Out Some Real World Separable Equations

In the following sections, I take a look at some real world examples featuring separable equations.

## Getting in control with a sample flow problem

To understand the relevance of differential equations in the real world, here's a sample problem to ponder: Say that you have a 10-liter pitcher of water, and that you're mixing juice concentrate into the pitcher at the same time that you're pouring juice out. If the concentrate going into the pitcher has $1/4$ kg of sugar per liter, the rate at which the concentrate is going into the pitcher, which I'll call $r_{in}$, is $1/100$ liter per second, and the juice in the pitcher starts off with 4 kg of sugar, find the amount of sugar in the juice, $Q$, as a function of time, $t$.

Because this problem involves a rate — $dQ/dt$, which is the change in the amount of sugar in the pitcher — it's a differential equation, not just a simple algebraic equation. I walk you through the steps of solving the equation in the following sections.

### Determining the basic numbers

When you start trying to work out this problem, remember that the change in the amount of sugar in the pitcher, $dQ/dt$, has to be the rate of sugar flow in minus the rate of sugar flow out, or something like this:

$$\frac{dQ}{dt} = (\text{rate of sugar flow in}) - (\text{rate of sugar flow out})$$

Now you ask: What's the rate of sugar flow in? That's easy; it's just the concentration of sugar in the juice concentrate multiplied by the rate at which the juice concentrate is flowing into the pitcher, which I'll call $r_{in}$. So, your equation looks something like this:

$$(\text{rate of sugar flow in}) = \frac{r_{in}}{4} \text{ kg/sec}$$

Now what about the flow of sugar out? The rate of sugar flow out is related to the rate at which juice leaves the pitcher. So if you assume that the amount of juice in the pitcher is constant, then $r_{in} = r_{out} = r$. That, in turn, means the rate of sugar flow out is the rate at which the juice leaves the pitcher, *r*, multiplied by the concentration of sugar in the juice, which is *Q* divided by the capacity of the pitcher (10 liters), or *Q*/10. Here's what your equation would look like:

$$\left(\text{rate of sugar flow out}\right) = \frac{Q}{10} r \text{ kg/sec}$$

So that means:

$$\frac{dQ}{dt} = \left(\text{rate of sugar flow in}\right) - \left(\text{rate of sugar flow out}\right) = \frac{r}{4} - \frac{Qr}{10}$$

where the initial condition is:

$$Q_0 = 4 \text{ kg}$$

### Solving the equation

The equation at the end of the previous section is separable, and separating the variables, each on their own side, gives you this equation:

$$\frac{dQ}{dt} + \frac{Qr}{10} = \frac{r}{4}$$

Now that, you might say, is a linear differential in *Q*. And you'd be right. So you know that the equation is both linear and separable.

You can handle this differential equation using the methods in Chapter 2. For instance, to solve, you find an integrating factor, multiply both sides by the integrating factor, and then see if you can figure out what product the left side is the derivative of and integrate it. Whew! It sounds rough, but note that the equation is of the following form:

$$\frac{dy}{dt} + ay = b$$

The solution to this kind of differential equation is already found in Chapter 2; you use an integrating factor of $e^{at}$. The solution to this kind of equation is:

$$y = (b/a) + ce^{-at}$$

So you can see that the solution to the juice flow problem is:

$$Q(t) = 2.5 + ce^{-rt/10}$$

Because $r = 1/100$ liter per second, the equation becomes:

$$Q(t) = 2.5 + ce^{-t/1000}$$

And because the initial condition is:

$$Q_0 = 4 \text{ kg}$$

you know that:

$$Q = 2.5 + 1.5\,e^{-t/1000}$$

Note the solution as $t \to \infty$ is 2.5 kg of sugar, and that's what you'd expect. Why? Because the concentrate has ¼ kg of sugar per liter, and 10 liters of water are in the pitcher. So 10/4 = 2.5 kg.

The direction field for different values of $Q_0$ appears in Figure 3-5. Notice that all the solutions tend toward the final $Q$ of 2.5 kg of sugar, as you'd expect.

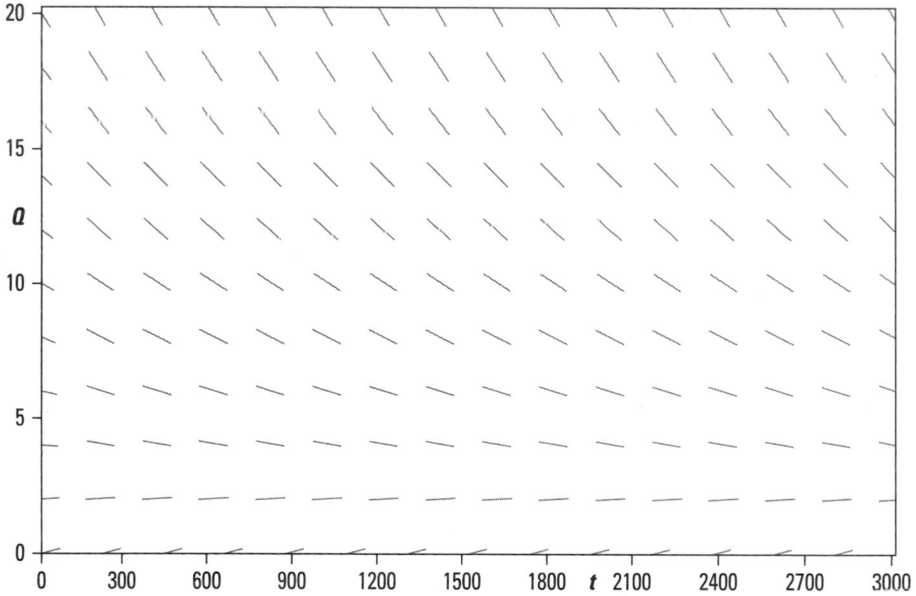


**Figure 3-5:** The direction field of a flow problem solution.

You can see a graph of this solution in Figure 3-6.

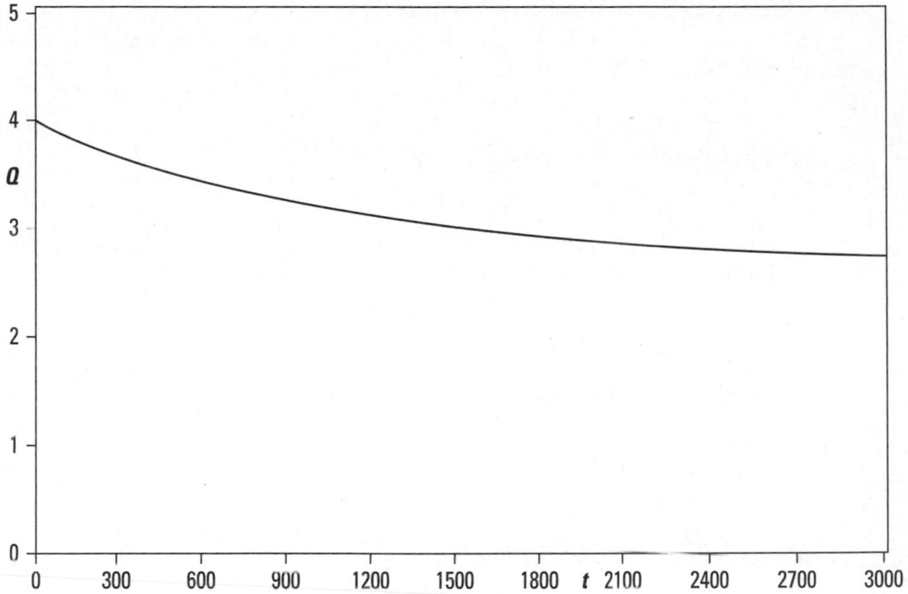


**Figure 3-6:** The graph of the solution of a flow problem.

## Striking it rich with a sample monetary problem

You may not have realized that differential equations can be used to solve money problems. Well they can! And here's a problem to prove it: Say that you're deciding whether to deposit your money in the bank. You can calculate how your money grows, $dQ/dt$, given the interest rate of the bank and the amount of money, $Q$, that you have in the bank. As you can see, this is a job for differential equations.

### Figuring out the general solution

Suppose your bank compounds interest continuously. The rate at which your savings, $Q$, grows, is:

$$\frac{dQ}{dt} = rQ$$

where $r$ is the interest rate that your bank pays.

REMEMBER

This equation says that the rate at which your money grows is equal to the interest rate multiplied by the current amount of money you have. That's an equation for the *rate* at which your money grows, not the actual amount of money.

Say that you have $Q_0$ money at $t = 0$:

$$Q(0) = Q_0$$

How much money would you have at a certain time in the future? That's easy enough to figure out. Separate the variables, each on their own side, like this:

$$\frac{dQ}{Q} = r\ dt$$

Then integrate:

$$\ln|Q| = rt$$

Finally, exponentiate both sides, which gives you the following equation:

$$Q = ce^{rt}$$

To match the initial condition:

$$Q(0) = Q_0$$

the solution becomes:

$$Q = Q_0e^{rt}$$

So, in other words, your money would grow exponentially. Not bad.

### Compounding interest at set intervals

Now I want you to examine the result from the previous section a little, deriving it another way so that it makes more sense. If your bank compounded interest once a year, not continuously, after $t$ years, you'd have this much money:

$$Q = Q_0(1 + r)^t$$

That's because if your interest was 5 percent, after the first year, you would have $1.05Q_0$; at the end of the second year, $1.05^2Q_0$, and so on.

What if your bank compounded interest twice a year? Would you have this much at the end of $t$ years:

$$Q = Q_0(1 + r)^{2t}$$

No, you wouldn't. Why? Because that would pay you $r$ percent interest twice a year. For example, if $r = 8$ percent, the previous equation would pay you 8 percent of your total savings twice a year. Instead, banks divide the interest rate they pay you by the number of times they compound per year, like this:

$$Q = Q_0\left(1 + \frac{r}{2}\right)^{2t}$$

In other words, if the bank compounds twice a year, and the annual interest rate is 8 percent, six months into the year it pays you 4 percent, and at the end of the year it pays another 4 percent.

In general, if your bank compounds interest $m$ times a year, after $t$ years, you'd have:

$$Q = Q_0\left(1 + \frac{r}{m}\right)^{mt}$$

If you take the limit as $m \to \infty$ — that is, as your bank starts to compound continuously — you get this equation:

$$Q = \lim_{m \to \infty} Q_0\left(1 + \frac{r}{m}\right)^{mt}$$

But that's just the expansion for $e^{rt}$. So, as the bank compounds continuously, you get:

$$Q = \lim_{m \to \infty} Q_0\left(1 + \frac{r}{m}\right)^{mt} = Q_0 e^{rt}$$

And this result confirms the answer you got from solving the differential equation in the previous section.

So if you had \$25 invested, and you left it alone at 6 percent for 60 years, you'd have:

$$Q = Q_0 e^{rt} = 25e^{0.06(60)}$$

or:

$$Q = Q_0 e^{rt} = 25e^{0.06(60)} = \$914.96$$

Hmm, not such a magnificent fortune.

### *Adding a set amount of money*

How about if you add a set amount every year to the equation in the previous section? That would be better, wouldn't it? Say that you add \$5,000 a year. In that case, remember that the set amount would change the differential equation for your savings, which was this:

$$\frac{dQ}{dt} = rQ$$

The equation would change to this, where $k$ is the amount you contribute regularly:

$$\frac{dQ}{dt} = rQ + k$$

If you deposit regularly, $k > 0$; if you withdraw regularly, $k < 0$. Ideally, you should add or subtract $k$ from your account continuously over the year to make your solution exact, but here you can just assume that you add or subtract $k$ once a year.

Putting this new equation into standard separable form gives you this:

$$\frac{dQ}{dt} - rQ = k$$

This equation is of the following form:

$$\frac{dy}{dt} + ay = b$$

The solution to this kind of equation is:

$$y = (b/a) + ce^{-at}$$

In this case, that solution means:

$$Q = ce^{rt} - k/r$$

What's going on here? It looks like you have the solution for leaving money in the bank without adding anything *minus* the amount you've added. Can that be right? The answer is in $c$, the constant of integration. Here, the initial condition is:

$$Q(0) = Q_0$$

which means that:

$$Q(0) = ce^{r0} - k/r = c - k/r = Q_0$$

or to simplify things:

$$c = Q_0 + k/r$$

So your solution turns out to be:

$$Q = ce^{rt} - k/r = (Q_0 + k/r)e^{rt} - k/r$$

Working this out gives you:

$$Q = \left(Q_0 + k/r\right)e^{rt} - k/r = Q_0 e^{rt} + \frac{k}{r}(e^{rt} - 1)$$

That looks a little better! Now the first term is the amount that you'd earn if you just left $Q_0$ in the account, and the second term is the amount resulting from depositing or withdrawing $k$ dollars regularly.

For example, say you started off with $25, but then you added $5,000 every year for 60 years. At the end of 60 years at 6 percent, you'd have:

$$Q = Q_0 e^{rt} + \frac{k}{r}(e^{rt} - 1) = 25e^{0.06\,(60)} + \frac{5,000}{0.06}(e^{0.06\,(60)} - 1)$$

After calculating this out, you'd get:

$$Q = 25e^{0.06\,(60)} + \frac{5,000}{0.06}(e^{0.06\,(60)} - 1) = \$914.96 + \$2,966,519 = \$2,967,434$$

Quite a tidy sum.

# Break It Up! Using Partial Fractions in Separable Equations

When a term in a separable differential equation looks a little difficult to integrate, you can use the *method of partial fractions* to separate it. This method is used to reduce the degree of the denominator of a rational expression.

For example, using the method of partial fractions, you can express:

$$\frac{6}{x^2 + 2x - 8}$$

as the following equation:

$$\frac{6}{x^2 + 2x - 8} = \frac{1}{x - 2} - \frac{1}{x + 4}$$

$X+4 - X-2 = 6$ ✓

**TIP**

Note that the power of the denominator has been reduced by one. You'll often see the method of partial fractions used when solving differential equations involving fractions, because using this method makes integrating the resulting two terms a lot easier.

Here's an example:

$$\frac{dy}{dx} = \frac{2xy}{x^2 - y^2}$$

You're probably screaming at your book right now. That's not separable, you say. However, perhaps you've also noticed that this equation is of the following form:

$$\frac{dy}{dx} = f(x, y)$$

And $f(x, y) = f(tx, ty)$, where $t$ is a constant. So you now have this equation:

$$\frac{dy}{dx} = \frac{2txty}{t^2x^2 - t^2y^2} = \frac{2xy}{x^2 - y^2}$$

**REMEMBER**

Yep, you guessed it: This equation calls for the old trick of substituting $y = vx$. (See the earlier section "A neat trick: Turning nonlinear separable equations into linear separable equations" for details.) Substituting $y = vx$ into this differential equation gives you the following equation:

$$v + x\frac{dv}{dx} = \frac{2x(xv)}{x^2 - (xv)^2}$$

or to simplify matters:

$$v + x\frac{dv}{dx} = \frac{2v}{1 - v^2}$$

$$\frac{v \cdot v}{1 - v^2} \quad = \quad \frac{v - v^3}{1 - v^2}$$

By subtracting the term on the right and adding to the left, you get the following:

$$x\frac{dv}{dx} + \frac{v(v^2 + 1)}{(v^2 - 1)} = 0$$

Flipping the fractions gives you:

$$\frac{dx}{x} + \frac{v^2 - 1}{v(v^2 + 1)}dv = 0$$

Time to use the method of partial fractions. In this case, you get the following equation:

$$\frac{dx}{x} + \frac{v^2 - 1}{v\left(v^2 + 1\right)}\, dv = \frac{dx}{x} + \left(\frac{2v}{v^2 + 1} - \frac{1}{v}\right) dv = 0$$

or to simplify:

$$\frac{dx}{x} + \frac{2v\, dv}{v^2 + 1} - \frac{dv}{v} = 0$$

Ah, much better. Now you can integrate! Integrating all these terms gives you:

$$\ln|x| + \ln|v^2 + 1| - \ln|v| = c$$

or:

$$\ln|x| + \ln|v^2 + 1| = \ln|v| + c$$

Exponentiating both sides gives you this equation:

$$x(v^2 + 1) = kv$$

where $k = e^c$.

Substituting $v = y/x$ gives you the following:

$$\frac{x\left(y^2 + x^2\right)}{x^2} = \frac{ky}{x}$$

The previous equation can then be simplified to the implicit solution of the following:

$$y^2 + x^2 = ky$$

or to simplify even further:

$$y^2 - ky + x^2 = 0$$

Finally, you can solve using the quadratic equation to get the following explicit solution:

$$y = \frac{k \pm \sqrt{k^2 - 4x^2}}{2}$$

## Chapter 4

# Exploring Exact First Order Differential Equations and Euler's Method

### In This Chapter

- Understanding the fundamentals of exact differential equations
- Figuring out whether differential equations are exact
- Turning nonexact equations into exact equations with integrating factors
- Crunching numbers with Euler's method
- Digging into difference equations

Not all first order differential equations are linear (see Chapter 2) or separable (see Chapter 3). So sometimes you have to use other tools to solve first order differential equations. One of those tools is knowing how to solve exact differential equations. That's what I introduce you to in this chapter. And for those really intractable (a.k.a. difficult) differential equations, you also get an introduction to working with mathematical methods. In particular, you find out how to use Euler's method to approximate solutions to just about any differential equation (assuming it has a solution!). And you find out that Euler's method is a type of difference equation (don't worry; I explain everything).

## Exploring the Basics of Exact Differential Equations

One of the most powerful methods for working with differential equations is seeing whether they're exact, and if they are, you can tackle them. I explain the basics in the following sections.

## Defining exact differential equations

REMEMBER

To solve an exact differential equation, you have to find a function whose partial derivatives correspond to the terms in the differential equation. For example, assume that you have a differential equation of this form:

$$M(x,y) + N(x,y)\frac{dy}{dx} = 0$$

where $M$ and $N$ are functions.

Now suppose that you can find a function $f(x, y)$ such that the following equations are true:

$$\frac{\partial f(x,y)}{\partial x} = M(x,y)$$

$$\frac{\partial f(x,y)}{\partial y} = N(x,y)$$

TIP

Note that the previous two equations are partial derivatives with respect to $x$, so I include the symbol $\partial$. (Check out Chapter 1 for more about partial derivatives.)

The differential equation that you're trying to solve becomes:

$$\frac{\partial f(x,y)}{\partial x} + \frac{\partial f(x,y)}{\partial y}\frac{dy}{dx} = 0$$

which is equal to (***note:*** ordinary derivatives now):

$$\frac{df(x,y)}{dx} = 0$$

So the solution, after integration, is:

$$f(x, y) = c$$

The previous solution is at least the implicit solution; you may have to move things around a bit to get the actual explicit solution. (As I explain in Chapter 3, implicit solutions aren't in the form $y = f(x)$, but explicit solutions are.)

REMEMBER

Here's the point: If you can find a function $f$ such that a differential equation can be reduced to the following form:

$$\frac{df(x,y)}{dx} = 0$$

then the differential equation is said to be *exact*.

## Working out a typical exact differential equation

Exact differential equations can be tough to solve, so in this section, I provide a typical example so you can get the hang of it. Check out this differential equation:

$$4x + 2y^2 + 4xy\frac{dy}{dx} = 0$$

This equation isn't linear, and it also isn't separable. However, you may suspect that the equation is exact. So, if that's the case, how do you solve it?

You might note that this function has some interesting properties:

$$f(x, y) = 2x^2 + 2xy^2$$

In particular, note that $\partial f/\partial x$ (***note:*** this is the partial derivative with respect to $x$) is:

$$\frac{\partial f}{\partial x} = 4x + 2y^2$$

Those are the first two terms of the equation you're looking to solve. Also, note that

$$\frac{\partial f}{\partial y} = 4xy$$

which looks a lot like the third term in the original equation. So you could write the equation like this:

$$\frac{\partial f}{\partial x} + \frac{\partial f}{\partial y}\frac{dy}{dx} = 0$$

You're making progress! Note that because $y$ is a function of $f$, you can use the chain rule (see Chapter 3) and switch to ordinary derivatives, meaning that you can write the original equation like this:

$$\frac{df}{dx} = 0$$

where

$$f(x, y) = 2x^2 + 2xy^2$$

The ordinary derivative I just gave you is easy to integrate, so you get the following equation:

$$2x^2 + 2xy^2 = c$$

where *c* is a constant of integration. That's the implicit solution. Now you can divide by 2, absorbing that 2 into *c*, which gives you:

$$x^2 + xy^2 = c$$

So that's your implicit solution. But it's easy to solve for *y*, as you can see in this equation:

$$y = \sqrt{\frac{c - x^2}{x}}$$

# Determining Whether a Differential Equation Is Exact

As you can see from the previous section, knowing when a differential equation is exact is helpful. But just when is it exact? Keep reading to find out.

## Checking out a useful theorem

REMEMBER

How can you tell, in a systematic way, whether a differential equation is exact? You're in luck: It turns out that there's a handy theorem to determine whether a differential equation is exact, and here it is:

**If the functions $M$, $N$, $\partial M/\partial y$, and $\partial N/\partial x$ are continuous in a rectangle $R$, then this differential equation:**

$$\mathbf{M(x,y) + N(x,y)\frac{dy}{dx} = 0}$$

**is an exact differential equation in the rectangle $R$ if and only if:**

$$\mathbf{\frac{\partial M(x,y)}{\partial y} = \frac{\partial N(x,y)}{\partial x}}$$

**at every point in $R$.**

In other words, if you have the differential equation

$$M(x,y) + N(x,y)\,\frac{dy}{dx} = 0$$

there exists a function $f(x, y)$ in a rectangle $R$ such that:

$$\frac{\partial f(x,y)}{\partial x} = M(x,y)$$

and

$$\frac{\partial f(x,y)}{\partial y} = N(x,y)$$

if and only if:

$$\frac{\partial M(x,y)}{\partial y} = \frac{\partial N(x,y)}{\partial x}$$

in the rectangle $R$.

REMEMBER

So how do you solve exact differential equations? You have to solve the partial differential equation:

$$\frac{\partial f(x,y)}{\partial x} - M(x,y)$$

and

$$\frac{\partial f(x,y)}{\partial y} = N(x,y)$$

And if you can find it, the implicit solution is $f(x, y) = c$.

## Applying the theorem

So how about an example that shows how to apply the theorem in the previous section? Take a look at this differential equation:

$$2xy + (1 + x^2)\,\frac{dy}{dx} = 0$$

In other words:

$$M(x, y) = 2xy$$

and

$$N(x, y) = (1 + x^2)$$

Note that

$$\frac{\partial M(x,y)}{\partial y} = \frac{\partial N(x,y)}{\partial x} = 2x$$

So now you know that the differential equation is exact. And now you need to find $M(x, y)$ and $N(x, y)$ such that:

$$\frac{\partial f(x,y)}{\partial x} = M(x,y)$$

and

$$\frac{\partial f(x,y)}{\partial y} = N(x,y)$$

When you find $f(x, y)$, the solution to the original equation is

$$f(x, y) = c$$

Start by finding $f(x, y)$. Because

$$M(x, y) = 2xy$$

that means

$$\frac{\partial f(x,y)}{\partial x} = 2xy$$

Integrating both sides of the previous equation gives you:

$$f(x, y) = x^2y + g(y)$$

that depends only on $y$, not on $x$. So what's $g(y)$? You

and

$$N(x, y) = (1 + x^2)$$

or

$$\frac{\partial f(x,y)}{\partial y} = (1 + x^2)$$

Because you know that

$$f(x, y) = x^2y + g(y)$$

you get this equation:

$$\frac{\partial f(x,y)}{\partial y} = x^2 + \frac{\partial g}{\partial y}(y) = 1 + x^2$$

By canceling out the $x^2$, you get

$$\frac{\partial g}{\partial y}(y) = 1$$

Fortunately, this equation is easy enough to integrate. Integrating it gives you this:

$$g(y) = y + d$$

where $d$ is a constant of integration. And because

$$f(x, y) = x^2y + g(y)$$

you get:

$$f(x, y) = x^2y + y + d$$

Whew. You now know that the solution to $f(x, y) = x^2y + g(y)$ is:

$$f(x, y) = c$$

So

$$x^2y + y = c$$

where the constant of integration $d$ has been absorbed into the constant of integration $c$.

As you recall from the earlier section "Defining exact differential equations," finding $f(x, y)$ gives you the implicit solution to the exact differential solution you're trying to solve. Happily, $x^2y + y = c$ is easy to solve for $y$ in terms of $x$, giving you:

$$y = \frac{c}{(1 + x^2)}$$

And that, my friends, is the explicit solution to your original equation. Cool!

# Conquering Nonexact Differential Equations with Integrating Factors

The earlier sections in this chapter cover exact differential equations. But what about those equations that don't look exact but can be converted into exact equations? Yes, you read that right! Sometimes, differential equations are intractable and not exact. For example, that's the case with this differential equation:

$$y - x\,\frac{dy}{dx} = 0$$

You start off by checking this differential equation for exactness, meaning that you have to cast it in this form:

$$M(x,y) + N(x,y)\,\frac{dy}{dx} = 0$$

So, in the case of the example:

$$M(x, y) = y$$

and

$$N(x, y) = -x$$

The differential equation is exact if:

$$\frac{\partial M(x,y)}{\partial y} = \frac{\partial N(x,y)}{\partial x}$$

But these two aren't equal; rather:

$$\frac{\partial M(x,y)}{\partial y} = 1$$

and

$$\frac{\partial N(x,y)}{\partial x} = -1$$

As you can see, the original equation isn't exact. However, the two partial derivatives, $\partial M/\partial x$ and $\partial N/\partial y$, differ only by a minus sign. Isn't there some way of making this differential equation exact?

Yup, you're right. There is a way! You just have to use an integrating factor. As I discuss in Chapter 2, integrating factors are used to multiply differential equations to make them easier to solve. In the following sections, I explain how to use an integrating factor to magically turn a nonexact equation into an exact one.

## Finding an integrating factor

How can you find an integrating factor that makes differential equations exact? Just follow the steps in the following sections.

### Multiplying by the factor you want to find

Using the example from earlier, here I show you how to find an integrating factor that makes differential equations exact. Say you have this differential equation:

$$M(x,y) + N(x,y)\frac{dy}{dx} = 0$$

Multiplying this equation by the integrating factor $\mu(x, y)$ (which you want to find) gives you this:

$$\mu(x,y)\,M(x,y) + \mu(x,y)\,N(x,y)\frac{dy}{dx} = 0$$

This equation is exact if:

$$\frac{\partial\big(\mu(x,y)\,M(x,y)\big)}{\partial y} = \frac{\partial\big(\mu(x,y)\,N(x,y)\big)}{\partial x}$$

which means that $\mu(x, y)$ must satisfy this differential equation:

$$M(x,y)\frac{\partial\mu(x,y)}{\partial y} - N(x,y)\frac{\partial\mu(x,y)}{\partial x} + \left[\frac{\big(\partial M(x,y)\big)}{\partial y} - \frac{\partial N(x,y)}{\partial x}\right]\mu(x,y) = 0$$

Well, yipes. That doesn't seem to have bought you much simplicity. In fact, this equation looks more complex than before. What you have to do now is to assume that $\mu(x, y)$ is a function of $x$ only — that is, $\mu(x, y) = \mu(x)$ — which also means that:

$$\frac{\partial \mu(x)}{\partial y} = 0$$

This turns the lengthy and complex equation into the following:

$$-N(x,y)\,\frac{\partial \mu(x)}{\partial x} + \left(\frac{\partial M(x,y)}{\partial y} - \frac{\partial N(x,y)}{\partial x}\right)\mu(x) = 0$$

Or in other words:

$$N(x,y)\,\frac{\partial \mu(x)}{\partial x} = \left(\frac{\partial M(x,y)}{\partial y} - \frac{\partial N(x,y)}{\partial x}\right)\mu(x)$$

Dividing both sides by $N(x, y)$ to simplify means that:

$$\frac{\partial \mu(x)}{\partial x} = \frac{1}{N(x,y)}\left(\frac{\partial M(x,y)}{\partial y} - \frac{\partial N(x,y)}{\partial x}\right)\mu(x)$$

Well, that looks somewhat better. You can see how to finish the example in the next section.

### *Completing the process*

Remember the differential equation that you're trying to solve? If not, here it is:

$$y - x\,\frac{dy}{dx} = 0$$

In this case,

$$M(x, y) = y$$

and

$$N(x, y) = -x$$

Plug these into your simplified equation from the previous section, which becomes:

$$\frac{\partial \mu(x)}{\partial x} = \frac{-1}{x}\left(1 - (-1)\right)\mu(x)$$

or:

$$\frac{\partial\mu(x)}{\partial x} = \frac{-2}{x}\mu(x)$$

Now this equation can be rearranged to get this one:

$$\frac{\partial\mu(x)}{\mu(x)} = -2\frac{\partial x}{x}$$

Integrating both sides gives you:

$$\ln|>(x)| = -2\ln|x|$$

And finally, exponentiating both sides results in this equation:

$$\pm\mu(x) = \frac{1}{x^2}$$

For this differential equation:

$$\frac{\partial M(x,y)}{\partial y} = 1$$

and

$$\frac{\partial N(x,y)}{\partial x} = -1$$

Because these equations differ by a sign, you may suspect that you need the negative version of $\mu(x)$, meaning that:

$$\mu(x) = -\frac{1}{x^2}$$

So that, at last, is your integrating factor. Whew!

## Using an integrating factor to get an exact equation

To carry on from the example in the previous section: Multiplying your original equation by your integrating factor gives you this new differential equation:

$$\frac{-y}{x^2} + \frac{1}{x}\frac{dy}{dx} = 0$$

Is this exact? To find out, consider that

$$M(x,y) = \frac{-y}{x^2}$$

and

$$N(x,y) = \frac{1}{x}$$

The differential equation is exact if:

$$\frac{\partial M(x,y)}{\partial y} = \frac{\partial N(x,y)}{\partial x}$$

You can now see that these two terms are equal, because:

$$\frac{\partial M(x,y)}{\partial y} = \frac{-1}{x^2}$$

and

$$\frac{\partial N(x,y)}{\partial x} = \frac{-1}{x^2}$$

So the integrating factor, $-1/x^2$, did the trick, and the equation is now exact.

## The finishing touch: Solving the exact equation

The last step is to solve the exact equation as I explain earlier in this chapter. In other words, you have to solve it by finding a function $f(x, y) = c$ such that:

$$\frac{\partial f(x,y)}{\partial x} = M(x,y)$$

and

$$\frac{\partial f(x,y)}{\partial y} = N(x,y)$$

As noted in the previous section,

$$M(x,y) = -\frac{y}{x^2}$$

which means that

$$\frac{\partial f(x,y)}{\partial x} = -\frac{y}{x^2}$$

Integrating both sides of the equation gives you:

$$f(x,y) = \frac{y}{x} + g(y)$$

And because, using the info from the previous section:

$$\frac{\partial f(x,y)}{\partial y} = N(x,y) = \frac{1}{x}$$

you can see that

$$g(y) = d$$

where $d$ is a constant of integration. So $f(x, y)$ must be:

$$f(x,y) = \frac{y}{x} + d$$

As you recall from the earlier section "Defining exact differential equations," the implicit solution of an exact differential equation is:

$$f(x, y) = c$$

So:

$$\frac{y}{x} = c$$

where the constant $d$ has been absorbed into the constant $c$. Now you know that the explicit solution to the exact equation is:

$$y = cx$$

Cool. The solution turned out to be simple.

# Getting Numerical with Euler's Method

Face it: Sometimes, you just can't find a solution to some differential equations. Or, at best, you may only be able to find an implicit one. So what do you do? In that case, it could be time to turn to numerical methods — that is, using a computer.

I know, using a computer feels a lot like a cop-out, but sometimes you have no other choice. Chapters 1, 2, and 3 all contain examples where I used a computer to calculate direction fields and plot some actual solutions. Now, in the following sections, I introduce one simple numerical method of solving differential equations. This method is called *Euler's method.*

## So who was this Euler guy?

Leonhard Paul Euler (1707–1783) was a Swiss mathematician who lived, for the most part, in Russia and Germany. He was an important genius who contributed to physics, calculus, and graph theory. He's also responsible for much of the mathematical terminology that people take for granted, such as the notation used for mathematical functions.

## *Understanding the method*

Euler's method basically says "Hey, you may not have the actual curve that represents the solution to your differential equation, but you have the slope of that curve everywhere (because the *slope* — the rate of change of the curve — is the derivative)."

To better understand what I mean, say that you have the following general differential equation:

$$\frac{dy}{dx} = f(x,y)$$

And suppose that you have a point, $(x_0, y_0)$, that's on the solution curve. Because of your differential equation, you know that the slope of the solution curve at that point is $f(x_0, y_0)$.

Now imagine that you want to find the solution at a point, $(x, y)$, a short distance away. Here's how you might find $y$:

$$y = y_0 + \Delta y$$

Anything with the symbol Δ signifies change. So, $\Delta y$ means a change in $y$.

Because the slope, $m$, is defined as $\Delta y/\Delta x$, this equals (for small $\Delta x$):

$$y = y_0 + m\,\Delta x$$

And because $\Delta x = x - x_0$, you have:

$$y = y_0 + m\,(x - x_0)$$

Because the slope, $m$, is equal to the derivative at $(x_0, y_0)$, and because of your original equation, $m = f(x_0, y_0)$, you get:

$$y = y_0 + f(x_0, y_0)\,(x - x_0)$$

So if you keep $(x - x_0)$ small, this approximation should be close. All it's doing is using the slope to extrapolate $(x_0, y_0)$ to a new point nearby, $(x, y)$. This process is illustrated in Figure 4-1.

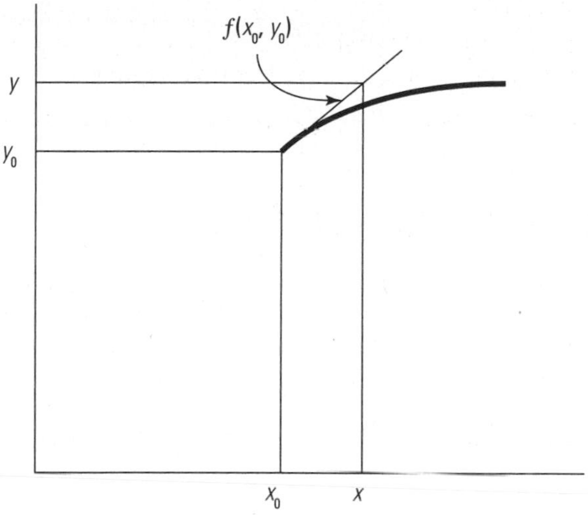


**Figure 4-1:** Euler's method at work.

TIP

If you call $\Delta x$ by the term $h$, as it's often referred to when using Euler's method, you get this:

$$y_1 = y_0 + f(x_0, y_0)\, h$$

REMEMBER

And in general, you can find any point along the solution curve like this, when using Euler's method:

$$y_{n+1} = y_n + f(x_n, y_n)\, h$$

## Checking the method's accuracy on a computer

In this section, I put Euler's method to work, finding a solution to the following differential equation:

$$\frac{dy}{dx} = x$$

where

$$y(0) = 0$$

In fact, you already know the solution by simply integrating:

$$y = \frac{x^2}{2}$$

Because you know the solution, you can now check the accuracy of Euler's method. And you're in luck, because here I show you how to develop a short program that uses Euler's method to solve differential equations. I developed this program using the Java programming language (which you can get for free at `java.sun.com`; just click the Java SE link under the Downloads tab).

You don't have to know Java to use this book — the programming in this section is only a demonstration. But because numerical methods of solving differential equations require a computer, I use a programming language to get things working. You can skip this part if you aren't interested.

### *Defining initial conditions and functions*

The Java program starts by defining the initial condition point — which is a solution point for the differential equation, by definition — $(x_0, y_0)$. Then it defines the step size, $h$, and the number of steps you want to take, $n$. So, for example, if you have the initial point at (0, 0), and you want the solution at (10, $y$), your step size is 0.1, and you need $n = 100$ steps:

```
double x0 = 0.0;
double y0 = 0.0;
double h = 0.1;
double n = 100;
```

At this point, the program also defines a Java function, $f(x, y)$, that returns the value of the derivative at any point $(x, y)$:

```
public double f(double x, double y)
{
  return x;
}
```

Because you know the exact solution, you'll also set up a Java function to return the exact solution at any point so you can compare it to the Euler's method value:

```
public double exact(double x, double y)
{
  return x * x / 2;
}
```

The actual work is done in a Java function named "calculate." In this function, each step using Euler's method is calculated and displayed, like this:

```
public void calculate()
{
  double x = x0;
  double y = y0;
  double k;

  System.out.println("x\t\tEuler\t\tExact");

  for (int i = 1; i < n; i++){
    k = f(x, y);
    y = y + h * k;
    x = x + h;
    System.out.println(round(x) + "\t\t" + round(y) +
         "\t\t" + round(exact(x, 0)));
  }

}
```

The program also has a function named "round," which rounds values to two decimal places for the printout:

```
public double round(double val)
{
  double divider = 100;
  val = val * divider;
  double temp = Math.round(val);
  return (double)temp / divider;
}
```

### *Examining the entire code*

The following is the whole code, e.java, which is a short program for calculating differential equation solutions using Euler's method. The parts you have to change when you want to solve your own differential equations are in **bold:**

```
public class e
{

  double x0 = 0.0;
  double y0 = 0.0;
  double h = 0.1;
  double n = 100;

  public e()
  {
```

```
    }

    public double f(double x, double y)
    {
      return x;
    }

    public double exact(double x, double y)
    {
      return x * x / 2;
    }

    public static void main(String [] argv)
    {
      e de = new e();
      de.calculate();
    }

    public void calculate()
    {
      double x = x0;
      double y = y0;
      double k;

      System.out.println("x\t\tEuler\t\tExact");

      for (int i = 1; i < n; i++){
        k = f(x, y);
        y = y + h * k;
        x = x + h;
        System.out.println(round(x) + "\t\t" + round(y) +
             "\t\t" + round(exact(x, 0)));
      }

    }

    public double round(double val)
    {
      double divider = 100;
      val = val * divider;
      double temp = Math.round(val);
      return (double)temp / divider;
    }
  }
```

### An example at work

In this section, I put the program to work. As it stands, e.java is set up with the following differential equation:

$$\frac{dy}{dx} = x$$

where

$$y(0) = 0$$

The program starts at $(x_0, y_0) = (0, 0)$, which you know is a point on the solution curve, and then it calculates 100 steps, each of $\Delta x = 0.1$.

You start by using the Java compiler, javac.exe, to compile the code:

```
C:\>javac e.java
```

Then, you run the compiled code, e.class, using java.exe like this: java e. The program then prints out the current $x$ value, the Euler approximation of the solution at that $x$ value, and the exact solution, as you see here:

```
C:\>java e
x                Euler            Exact
0.1              0.0              0.01
0.2              0.01             0.02
0.3              0.03             0.05
0.4              0.06             0.08
0.5              0.1              0.13
0.6              0.15             0.18
0.7              0.21             0.24
0.8              0.28             0.32
0.9              0.36             0.4
1.0              0.45             0.5
1.1              0.55             0.6
1.2              0.66             0.72
1.3              0.78             0.85
1.4              0.91             0.98
1.5              1.05             1.13
1.6              1.2              1.28
1.7              1.36             1.45
1.8              1.53             1.62
1.9              1.71             1.81
2.0              1.9              2.0
2.1              2.1              2.21
2.2              2.31             2.42
2.3              2.53             2.65
2.4              2.76             2.88
2.5              3.0              3.13
2.6              3.25             3.38
2.7              3.51             3.65
2.8              3.78             3.92
2.9              4.06             4.21
3.0              4.35             4.5
3.1              4.65             4.81
3.2              4.96             5.12
3.3              5.28             5.45
```

| | | |
|---|---|---|
| 3.4 | 5.61 | 5.78 |
| 3.5 | 5.95 | 6.13 |
| 3.6 | 6.3 | 6.48 |
| 3.7 | 6.66 | 6.85 |
| 3.8 | 7.03 | 7.22 |
| 3.9 | 7.41 | 7.61 |
| 4.0 | 7.8 | 8.0 |
| 4.1 | 8.2 | 8.41 |
| 4.2 | 8.61 | 8.82 |
| 4.3 | 9.03 | 9.25 |
| 4.4 | 9.46 | 9.68 |
| 4.5 | 9.9 | 10.13 |
| 4.6 | 10.35 | 10.58 |
| 4.7 | 10.81 | 11.04 |
| 4.8 | 11.28 | 11.52 |
| 4.9 | 11.76 | 12.0 |
| 5.0 | 12.25 | 12.5 |
| 5.1 | 12.75 | 13.0 |
| 5.2 | 13.26 | 13.52 |
| 5.3 | 13.78 | 14.04 |
| 5.4 | 14.31 | 14.58 |
| 5.5 | 14.85 | 15.12 |
| 5.6 | 15.4 | 15.68 |
| 5.7 | 15.96 | 16.24 |
| 5.8 | 16.53 | 16.82 |
| 5.9 | 17.11 | 17.4 |
| 6.0 | 17.7 | 18.0 |
| 6.1 | 18.3 | 18.6 |
| 6.2 | 18.91 | 19.22 |
| 6.3 | 19.53 | 19.84 |
| 6.4 | 20.16 | 20.48 |
| 6.5 | 20.8 | 21.12 |
| 6.6 | 21.45 | 21.78 |
| 6.7 | 22.11 | 22.44 |
| 6.8 | 22.78 | 23.12 |
| 6.9 | 23.46 | 23.8 |
| 7.0 | 24.15 | 24.5 |
| 7.1 | 24.85 | 25.2 |
| 7.2 | 25.56 | 25.92 |
| 7.3 | 26.28 | 26.64 |
| 7.4 | 27.01 | 27.38 |
| 7.5 | 27.75 | 28.12 |
| 7.6 | 28.5 | 28.88 |
| 7.7 | 29.26 | 29.64 |
| 7.8 | 30.03 | 30.42 |
| 7.9 | 30.81 | 31.2 |
| 8.0 | 31.6 | 32.0 |
| 8.1 | 32.4 | 32.8 |
| 8.2 | 33.21 | 33.62 |
| 8.3 | 34.03 | 34.44 |
| 8.4 | 34.86 | 35.28 |

| | | |
|---|---|---|
| 8.5 | 35.7 | 36.12 |
| 8.6 | 36.55 | 36.98 |
| 8.7 | 37.41 | 37.84 |
| 8.8 | 38.28 | 38.72 |
| 8.9 | 39.16 | 39.6 |
| 9.0 | 40.05 | 40.5 |
| 9.1 | 40.95 | 41.4 |
| 9.2 | 41.86 | 42.32 |
| 9.3 | 42.78 | 43.24 |
| 9.4 | 43.71 | 44.18 |
| 9.5 | 44.65 | 45.12 |
| 9.6 | 45.6 | 46.08 |
| 9.7 | 46.56 | 47.04 |
| 9.8 | 47.53 | 48.02 |
| 9.9 | 48.51 | 49.0 |

So there you have it — Euler's method looks pretty accurate for your differential equation's solution. At $x = 9.9$, it's off only by 1% (49.00 – 48.51). Not bad. ***Tip:*** You can get even better accuracy by using smaller step sizes.

WARNING!

You have to be careful when it comes to step size. If you have a really steep slope, you have to use a truly tiny step size to get the most accurate results.

# Delving into Difference Equations

Euler's method, which is discussed in the previous section, brings up an interesting topic: *difference equations.* That's correct, you read right — "difference" equations. I know what you're thinking: An explanation is in order! Allow me to start at the beginning. Derivatives are built to work like this, where $\Delta x \to 0$ and $y = f(x)$:

$$\frac{dy}{dx} = \lim_{\Delta x \to 0} \frac{f(x + \Delta x) - f(x)}{\Delta x}$$

However, sometimes, you don't want $\Delta x$ to go to zero. In other words, sometimes it just makes sense to keep the step size, $\Delta x$, discrete and nonzero. For example, you can set up a differential equation to calculate the interest you pay on a loan, but because $\Delta x \to 0$, the differential equation calculates the interest as if it were compounded continuously. But what if the interest is actually compounded monthly or annually? In that case, you should use $\Delta x$ = 1 month or 1 year, not $\Delta x \to 0$.

REMEMBER

When $\Delta x$ doesn't go to zero, you have a difference equation, not a differential equation. I take some time to look at difference equations in the following sections.

## Some handy terminology

Euler's method is a good example of a difference equation, because every $y$ value depends on the previous $y$ value:

$$y_{n+1} = y_n + f(x_n, y_n)\, h$$

Here's how it looks written as a difference equation, where the value $y_{n+1}$ is some function of $y_n$:

$$y_{n+1} = f(n, y_n)$$

***Note:*** This is the kind of equation you'd use if interest was compounded annually in a savings account instead of continuously. In that case, the future value of the savings account, $y_{n+1}$, depends on the current value, $y_n$.

REMEMBER

Just as with differential equations, you can apply some terminology to difference equations. In particular, the previous equation is a *first order difference equation.* Why? Because it depends on $y_n$, not on earlier terms ($y_{n-1}$, $y_{n-2}$, and so on). The equation is *linear* if $f(n, y_n)$ is linear in $y_n$. If $f(n, y_n)$ isn't linear in $y_n$, the equation is *nonlinear.*

You can also have initial conditions for difference equations, such as the first value being set to a constant:

$$y_0 = c$$

## Iterative solutions

Because a difference equation is defined using successive terms:

$$y_{n+1} = f(n, y_n)$$

the solution is all the terms $y_0, y_1, y_2 \ldots y_{n+1}$.

To make solving these kinds of equations a little more manageable, assume that the function $f$ in the original equation only depends on $y_n$, not on $n$ itself, so you get:

$$y_{n+1} = f(y_n)$$

So:

$$y_1 = f(y_0)$$

and:

$$y_2 = f(y_1)$$

which means that:

$$y_2 = f(y_1) = f(f(y_0))$$

This equation is sometimes written as:

$$y_2 = f(y_1) = f(f(y_0)) = f^2(y_0)$$

which is often called the second *iterate* of the difference equation's solution, written as $f^2(y_0)$.

Here's the third iterate:

$$y_3 = f(y_2) = f(f(y_1)) = f(f(f(y_0)))$$

The third iterate can also be written using the $f^n()$ notation, like this:

$$y_3 = f(y_2) = f(f(y_1)) = f(f(f(y_0))) = f^3(y_0)$$

REMEMBER

As long as $y_{n+1} = f(y_n)$, you can write the $n$th iterate as:

$$y_n = f^n(y_0)$$

## Equilibrium solutions

It's often important to know what happens as $n \to \infty$. Does the series converge? Does it diverge? The answer tells you whether you have a viable solution.

REMEMBER

Sometimes, all the $y_n$ have the same value. In this case, the solution is said to be an *equilibrium solution.* The series converges to that equilibrium solution. So, every term is the same:

$$y_n = f^n(y_0)$$

and:

$$y_n = f(y_n)$$

I introduce some examples in the following sections.

### Working without a constant

To better understand equilibrium solutions, say, for example, that you have savings in a bank that pays an interest rate, $i$, annually. The amount of money you have in year, $n + 1$, is written like so:

$$y_{n+1} = (1 + i)\, y_n$$

The solution of this difference equation is easy, because $i$ is constant. Each year's savings is simply $(1 + i)$ multiplied by the previous year's savings. So your equation looks like this:

$$y_n = (1 + i)^n\, y_0$$

This means that the limiting value as $n \to \infty$ depends on $i$.

For instance, if $i$ is less than 0, you get:

$$\lim_{n \to \infty} y_n = 0$$

If $i$ is equal to 0, you get:

$$\lim_{n \to \infty} y_n = y_0$$

Otherwise, you get:

$$\lim_{n \to \infty} y_n = \text{nonexistent}$$

That is to say, the equilibrium solution is $i = 0$.

### Working with a constant

Just for kicks, change the circumstances from the previous section. Say that you increase your savings each year by adding a constant value, $c$. Your equation would now look like this:

$$y_{n+1} = (1 + i)\, y_n + c$$

The solution to this difference equation is:

$$y_n = (1 + i)^n y_0 + (1 + (1 + i) + (1 + i)^2 + (1 + i)^{n-1})c$$

If $i$ isn't equal to 0, you can write the equation as:

$$y_n = (1 + i)^n y_0 + \frac{(1 + i)^n - 1}{i} c$$

For example, if you started with $y_0$ = \$1,000 and $i$ = 5%, and you add \$1,000 a year for 10 years, you'd end up with an equation that looks like this:

$$y_n = (1+i)^n y_0 + \frac{(1+i)^n - 1}{i} c = 1.62(1{,}000) + \frac{1.62-1}{.05}(1{,}000)$$

or:

$$y_n = 1.62(1{,}000) + \frac{1.62-1}{.05}(1{,}000) = 1{,}620 + 12{,}400 = \$14{,}020$$

Not a bad return!

# Part II

# Surveying Second and Higher Order Differential Equations

The 5th Wave By Rich Tennant

"We're both mathematicians, Sheldon, so let me explain it this way. Where *r* denotes the ordinate of our relationship at the time, *t*, above the point *x*, where *b* denotes boring, *o* denotes over, *m* denotes moving on...are you starting to get any of this?"

## In this part . . .

It's time to up the ante with second order — and higher — differential equations. A second order differential equation is one that involves a second derivative; higher order equations involve three or more.

In this part, you discover that there are dazzling new techniques to bring to bear with second and higher order equations, such as the popular method of undetermined coefficients and variation of parameters. Get set to exercise your brain!

# Chapter 5

# Examining Second Order Linear Homogeneous Differential Equations

*In This Chapter*

- Focusing on the fundamentals of second order linear differential equations
- Getting a grip on constant coefficients
- Exploring characteristic equations
- Taking a crack at reduction of order
- Understanding the Wronskian and other cool theorems

In Part I, I tell you what you need to know about a variety of first order differential equations. Now you're striking out into uncharted territory — second order differential equations. These kinds of equations are based heavily in physics — in wave motion, electromagnetic circuits, heat conduction, and so on. They're also fun and interesting. In this chapter, I walk you through the fundamentals of second order linear homogeneous differential equations, with a few useful tips and theorems thrown in along the way.

## The Basics of Second Order Differential Equations

To better understand second order differential equations, first take a look at the following equation:

$$\frac{d^2y}{dx^2} = f(x, y)$$

As you can see, this equation has a second derivative, which makes it a second order differential equation. But you know what? That's not quite good enough. The $f(x, y)$ is okay for first order differential equations, but it isn't good enough for second order ones because the function $f$ may also depend on $dy/dx$. So the following is the general form of a second order differential equation:

$$\frac{d^2y}{dx^2} = f\left(x, y, \frac{dy}{dx}\right)$$

In the following sections, I introduce several important types of second order differential equations, including linear equations and homogeneous equations.

In this chapter, I focus on the second order linear homogeneous differential equation (try saying that ten times fast!); if you can solve the homogeneous form of a differential equation, you can always solve the same differential equation as a nonhomogeneous differential equation (or at least give the solution in integral form). In other words, finding the solution to the homogeneous differential equation is the fundamental part of solving second order differential equations. (I show you how to solve second order linear nonhomogeneous differential equations in Chapter 6. As I note there, solving nonhomogeneous equations usually involves solving the corresponding homogeneous equation as well.)

## Linear equations

I restrict this chapter to second order linear differential equations that have the following form:

$$y'' + p(x)y' + q(x)y = g(x)$$

where:

$$y'' = \frac{d^2y}{dx^2}$$

and:

$$y' = \frac{dy}{dx}$$

However, a typical second order linear equation is sometimes written as the following (who says that mathematicians don't like to mix things up a bit?):

$$P(x)y'' + Q(x)y' + R(x)y = G(x)$$

where $P(x)$, $Q(x)$, $R(x)$, and $G(x)$ are functions. The transition between the original equation and the alternate is easy. Check it out:

$$p(x) = \frac{Q(x)}{P(x)} \qquad q(x) = \frac{R(x)}{P(x)} \qquad g(x) = \frac{G(x)}{P(x)}$$

In this chapter, I show you only how to solve the typical second order linear equation for regions where $p(x)$, $q(x)$, and $g(x)$ are *continuous functions,* meaning that their values don't make discontinuous jumps. I also usually provide initial conditions as well, such as:

$$y(x_0) = y_0$$

But that condition isn't enough to specify the solution of a second order linear differential equation. You also need to specify the value of $y'(x_0)$. That value is usually something like this:

$$y'(x_0) = y'_0$$

TIP

Second order differential equations that aren't in the form of $y'' + p(x)y' + q(x)y = g(x)$ are called *nonlinear.*

## Homogeneous equations

REMEMBER

The equation $y'' + p(x)y' + q(x)y = g(x)$ is referred to as *homogeneous* if $g(x) = 0$; that is, the equation is of the following form:

$$y'' + p(x)y' + q(x)y = 0$$

Alternately, using the $P(x)$, $Q(x)$, $R(x)$, $G(x)$ terminology from the previous section, a homogeneous equation can be written as:

$$P(x)\, y'' + Q(x)y' + R(x)y = 0$$

TIP

If second order linear differential equations can't be put into either of these forms, the equation is said to be *nonhomogeneous.* To find out how to solve linear nonhomogeneous differential equations, check out Chapter 6.

# Second Order Linear Homogeneous Equations with Constant Coefficients

You may think that second order linear homogeneous differential equations are intimidating. But they really aren't, if you know some fundamentals.

The best place to start solving second order differential equations is with equations where $P(x)$, $Q(x)$, and $R(x)$ are constants, $a$, $b$, and $c$. So, for example, you get this equation when you include the constants:

$$ay'' + by' + cy = 0$$

Okay, so you've narrowed things down. Now you're talking about *second order linear homogeneous differential equations with constant coefficients.* Despite the hairy name, these equations are pretty easy to solve. In fact, you can always solve an equation of this type using some elementary solutions and initial conditions. Take a look at the examples in the following sections.

## Elementary solutions

To get you up to speed, I begin with this typical equation:

$$y'' - y = 0$$

Yes, this qualifies as a second order linear homogeneous differential equation. Why? Just figure that $a = 1$, $b = 0$, and $c = -1$. There you have it.

To solve this differential equation, you need a solution $y = f(x)$ whose second derivative is the same as $f(x)$ itself, because subtracting the $f(x)$ from $f''(x)$ gives you 0.

You can probably think of one such solution: $y = e^x$. Substituting $y = e^x$ into the equation gives you this:

$$e^x - e^x = 0$$

As you can see, $y = e^x$ is indeed a solution.

In fact, $y = c_1e^x$ (where $c_1$ is a constant) is also a solution, because $y''$ still equals $c_1e^x$, which means that substituting $y = c_1e^x$ into the original equation gives you:

$$c_1e^x - c_1e^x = 0$$

So $y = c_1e^x$ is also a solution. In fact, that solution is more general than just $y = e^x$, because $y = c_1e^x$ represents an infinite number of solutions, depending on the value of $c_1$.

You can go further still by noting that $y = e^{-x}$ is also a solution, because:

$$y'' - y = e^{-x} - e^{-x} = 0$$

Again, note that if $y = e^{-x}$ is a solution, then $y = c_2e^{-x}$ (where $c_2$ is a constant) is also a solution because:

$$y'' - y = c_2e^{-x} - c_2e^{-x} = 0$$

And here's the fun part: If $y = c_1e^x$ and $y = c_2e^{-x}$ are both solutions, then the sum of these two must also be a solution:

$$y = c_1e^x + c_2e^{-x}$$

In other words, if $y = f_1(x)$ and $y = f_2(x)$ are solutions to a second order linear homogeneous differential equation, then:

$$y = f_1(x) + f_2(x)$$

is also a solution.

## Initial conditions

In this section, I give you a look at some initial conditions to fit with the solutions from the previous section. For instance, say that you have these conditions:

$$y(0) = 9$$

and

$$y'(0) = -1$$

To meet these initial conditions, you can use the form of the solution $y = c_1e^x + c_2e^{-x}$, which means that $y' = c_1e^x - c_2e^{-x}$. Using the initial conditions, you get:

$$y(0) = c_1e^x + c_2e^{-x} = c_1 + c_2 = 9$$

$$y'(0) = c_1e^x - c_2e^{-x} = c_1 - c_2 = -1$$

So you get these two equations:

$$c_1 + c_2 = 9$$

$$c_1 - c_2 = -1$$

These are two equations in two unknowns. To solve them, write $c_1 + c_2 = 9$ in this form:

$$c_2 = 9 - c_1$$

Now substitute this expression for $c_2$ into $c_1 - c_2 = -1$, which gives you this equation:

$$c_1 - 9 + c_1 = -1$$

or:

$$2c_1 = 8$$

So

$$c_1 = 4$$

Substituting the preceding value of $c_1$ into $c_1 + c_2 = 9$ gives you:

$$c_1 + c_2 = 4 + c_2 = 9$$

or:

$$c_2 = 5$$

These values of $c_1$ and $c_2$ give you the following solution:

$$y = 4e^x + 5e^{-x}$$

That was simple enough, and you can generalize it as the solution to any linear homogeneous second order differential equation with constant coefficients.

# Checking Out Characteristic Equations

Here's the general second order linear homogeneous equation, where $a$, $b$, and $c$ are constants:

$$ay'' + by' + cy = 0$$

The solution of this equation is of the form $y = e^{rx}$, where $r$ isn't yet determined. Plugging $y = e^{rx}$ into the equation gives you:

$$ar^2e^{rx} + bre^{rx} + ce^{rx} = 0$$

Dividing by $e^{rx}$ (which is always nonzero) gives you:

$$ar^2 + br + c = 0$$

This equation is called the *characteristic equation* for $ay'' + by' + cy = 0$, and it's a quadratic equation. If the roots of the characteristic equation are $r_1$ and $r_2$, the general solution of $ay'' + by' + cy = 0$ is:

$$y = c_1e^{r_1x} + c_2e^{r_2x}$$

Earlier in this chapter, I treat this equation as *one* solution of $ay'' + by' + cy = 0$. But the truth is that it's *the unique* solution, as is indicated by a theorem later in this chapter (see the later section "Putting Everything Together with Some Handy Theorems"). Because $ar^2 + br + c = 0$ is a quadratic equation, the following are three possibilities for $r_1$ and $r_2$:

- $r_1$ and $r_2$ are real and distinct
- $r_1$ and $r_2$ are complex numbers (complex conjugates of each other)
- $r_1 = r_2$, where $r_1$ and $r_2$ are real

You can take a look at these cases in the following sections.

## Real and distinct roots

When it comes to characteristic equations, one possibility is that $r_1$ and $r_2$ are real, and not equal to each other. I explain the basics of understanding this concept and provide a clarifying example in the following sections.

### The basics

When solving a general problem, where:

$$y(x_0) = y_0$$

and:

$$y'(x_0) = y'_0$$

you get the following:

$$y_0 = c_1 e^{r_1 x_0} + c_2 e^{r_2 x_0}$$

and

$$y'_0 = c_1 r_1 e^{r_1 x_0} + c_2 r_1 e^{r_2 x_0}$$

Then you can solve for $c_1$ and $c_2$ in these equations, which gives you:

$$c_1 = \frac{y'_0 - y_0 r_2}{r_1 - r_2} e^{-r_1 x_0}$$

and:

$$c_2 = \frac{y_0 r_1 - y'_0}{r_1 - r_2} e^{-r_2 x_0}$$

### An example

How about an example to bring the concept of real and distinct roots into focus? Try this second order linear homogeneous differential equation:

$$y'' + 5y' + 6y = 0$$

with the initial condition that:

$$y(0) = 16$$

and:

$$y'(0) = -38$$

To solve, make the assumption that the solution is of the form $y = ce^{rx}$. Substituting that equation into $y'' + 5y' + 6y = 0$ gives you:

$$cr^2e^{rx} + 5cre^{rx} + 6ce^{rx} = 0$$

Next you divide by $ce^{rx}$ to get:

$$r^2 + 5r + 6 = 0$$

which is the characteristic equation for $y'' + 5y' + 6y = 0$. You can also write this equation as:

$$(r + 2)(r + 3) = 0$$

So the roots of the characteristic equation are:

$$r_1 = -2$$

and

$$r_2 = -3$$

which means that the solution to $y'' + 5y' + 6y = 0$ is:

$$y = c_1e^{-2x} + c_2e^{-3x}$$

where $c_1$ and $c_2$ are determined by the initial conditions:

$$y(0) = 16$$

and:

$$y'(0) = -38$$

Substituting the initial conditions into your solution gives you these equations:

$$y(0) = c_1 + c_2 = 16$$

and:

$$y'(0) = -2c_1 - 3c_2 = -38$$

From the first equation, you can see that $c_2 = 16 - c_1$, and substituting that into the second equation gives you:

$$-2c_1 - 3c_2 = -2c_1 - 48 + 3c_1 = c_1 - 48 = -38$$

or:

$$c_1 = 10$$

Substituting this value of $c_1$ into $y(0) = c_1 + c_2 = 16$ gives you:

$$c_1 + c_2 = 10 + c_2 = 16$$

So that means:

$$c_2 = 6$$

Now you've found $c_1$ and $c_2$, which means that the general solution to $y'' + 5y' + 6y = 0$ is:

$$y = 10e^{-2x} + 6e^{-3x}$$

You can see this solution graphed in Figure 5-1.

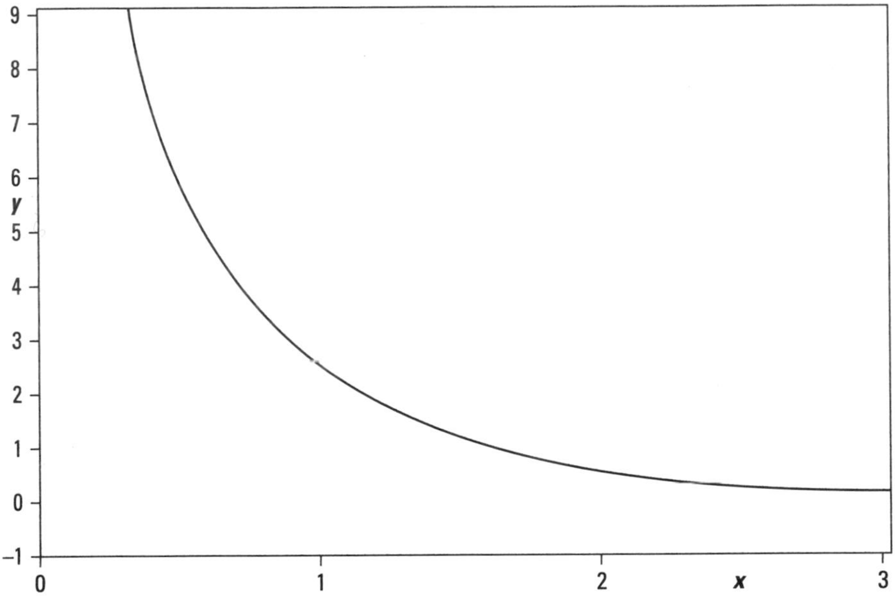


**Figure 5-1:** The solution to an equation with real and distinct roots.

## Complex roots

Besides real and distinct roots (see the previous section), another possibility, when it comes to characteristic equations, is having complex roots. In this case, the quadratic formula yields two complex numbers. I explain the basics of this concept and walk you through an example in the following sections.

### The basics

Check out the following quadratic equation:

$$r = \frac{-b \pm \sqrt{b^2 - 4ac}}{2a}$$

REMEMBER

The discriminant, $b^2 - 4ac$, is negative, which means that you're taking the square root of a negative number, so you get complex numbers. In particular, the imaginary part of the two roots varies by their sign (+ or –), as shown by the example equation.

In other words, the roots are of the following forms:

$$r_1 = \lambda + i\mu$$

and

$$r_2 = \lambda - i\mu$$

where $\lambda$ and $\mu$ are both real numbers, and $i$ is the square root of –1.

As you can see, the characteristic equation has complex roots. What does that mean for the solution to the differential equation? As you find out earlier in this chapter, the solutions to the differential equation are:

$$y_1 = e^{r_1 x}$$

and

$$y_2 = e^{r_2 x}$$

So:

$$y_1 = e^{(\lambda + i\mu)x}$$

and:

$$y_2 = e^{(\lambda - i\mu)x}$$

Now you have to deal with complex numbers as exponents of $e$. You may be familiar with the following handy equations:

$$e^{iax} = \cos ax + i \sin ax$$

and:

$$e^{-iax} = \cos ax - i \sin ax$$

You're making progress! You have now removed $i$ from the exponent. Putting these two equations to work gives you these forms for the solutions, $y_1$ and $y_2$:

$$y_1 = e^{(\lambda + i\mu)x} = e^{\lambda x}(\cos \mu x + i \sin \mu x)$$

and:

$$y_2 = e^{(\lambda - i\mu)x} = e^{\lambda x}(\cos \mu x - i \sin \mu x)$$

However, you still have that pesky factor of $i$; and you want to get rid of it. Why? The general differential equation that $y_1$ and $y_2$ are solutions of only has real coefficients:

$$ay'' + by' + cy = 0$$

Here's the key: You could get rid of those factors of $i$ if you could just divide them out. After all, $i$ is just a constant. For instance, if you had the solution in a form like this:

$$y_n = i\, e^{\lambda x} \cos \mu x$$

you could replace $i$ with an arbitrary constant of integration, $c$, which would look like this:

$$y_n = c\, e^{\lambda x} \cos \mu x$$

Now try adding and subtracting $y_1$ and $y_2$. As you recall from the earlier section "Elementary solutions," if two functions are a solution to a linear differential equation, the sum and difference of those functions are also solutions. So you can introduce the new solutions $m(x)$ and $n(x)$, the sum and difference of $y_1$ and $y_2$:

$$m(x) = y_1(x) + y_2(x)$$

and:

$$n(x) = y_1(x) - y_2(x)$$

First, calculating $m(x)$ gives you the following equation:

$$m(x) = y_1(x) + y_2(x) = e^{\lambda x}(\cos \mu x + i \sin \mu x) + e^{\lambda x}(\cos \mu x - i \sin \mu x)$$

which you can convert to:

$$m(x) = y_1(x) + y_2(x) = 2\, e^{\lambda x} \cos \mu x$$

Your solution now looks fine — there's no pesky $i$ lurking around. Now how about calculating $n(x)$? Here's what that looks like:

$$n(x) = y_1(x) - y_2(x) = e^{\lambda x}(\cos \mu x + i \sin \mu x) - e^{\lambda x}(\cos \mu x - i \sin \mu x)$$

which works out to:

$$n(x) = y_1(x) - y_2(x) = 2i\, e^{\lambda x} \sin \mu x$$

Your solution is looking even better, because now $m(x)$ and $n(x)$ have the following forms:

$$m(x) = c_1\, e^{\lambda x}\cos \mu x$$

and:

$$n(x) = c_2\, e^{\lambda x} \sin \mu x$$

Yes, it's true that $n(x)$ as written is multiplied by $i$: $2i\, e^{\lambda x} \sin \mu x$. But that's the beauty of the whole thing: $2i$ is just a constant, so it can be replaced by $c_2$. And that substitution gets rid of the pesky $i$.

So, finally, you can write the solution as:

$$y(x) = m(x) + n(x)$$

or:

$$y(x) = c_1\, e^{\lambda x} \cos \mu x + c_2\, e^{\lambda x} \sin \mu x$$

where you get $\lambda$ and $\mu$ from the roots of the differential equation's characteristic equation:

$$ar^2 + br + c = 0$$

where the roots are:

$$r_1 = \lambda + i\mu$$

and

$$r_2 = \lambda - i\mu$$

### An example

If you feel like you've grasped the basics of complex roots, take a look at an example. Try solving this differential equation:

$$2y'' + 2y' + y = 0$$

where:

$$y(0) = 1$$

and

$$y'(0) = 1$$

You probably already have a good idea what to do here. You simply need to find the characteristic equation and then the roots of that equation. Here's the characteristic equation:

$$2r^2 + 2r + 1 = 0$$

You can find the roots of this characteristic equation by using the quadratic equation from the previous section. Your work should give you these roots:

$$\frac{-2 \pm \sqrt{4-8}}{4}$$

This equation works out to be:

$$r = -\tfrac{1}{2} \pm (\tfrac{1}{2})i$$

So the two roots are:

$$r_1 = -\tfrac{1}{2} + (\tfrac{1}{2})i$$

and:

$$r_2 = -\tfrac{1}{2} - (\tfrac{1}{2})i$$

Because the two roots are of the form:

$$r_1 = \lambda + i\mu$$

and

$$r_2 = \lambda - i\mu$$

then:

$$\lambda = -\tfrac{1}{2}$$

and:

$$\mu = \tfrac{1}{2}$$

So the solution is:

$$y(x) = c_1\, e^{-x/2} \cos \tfrac{x}{2} + c_2\, e^{-x/2} \sin \tfrac{x}{2}$$

To find $c_1$ and $c_2$, you simply have to apply the initial conditions (substituting in for y(0) and y′(0)), which means that:

$$y(0) = c_1 = 1$$

and:

$$y'(0) = \frac{1}{2c_1} + \frac{1}{2c_2} = 1$$

Substituting $c_1 = 1$ gives you:

$$y'(0) = \frac{1}{2} + \frac{1}{2c_2} = 1$$

So $c_2 = 1$. That makes the general solution to $2y'' + 2y' + y = 0$:

$$y(x) = e^{-x/2} \cos \frac{x}{2} + e^{-x/2} \sin \frac{x}{2}$$

And that's it — no imaginary terms involved. Cool. You can see a graph of this solution in Figure 5-2.

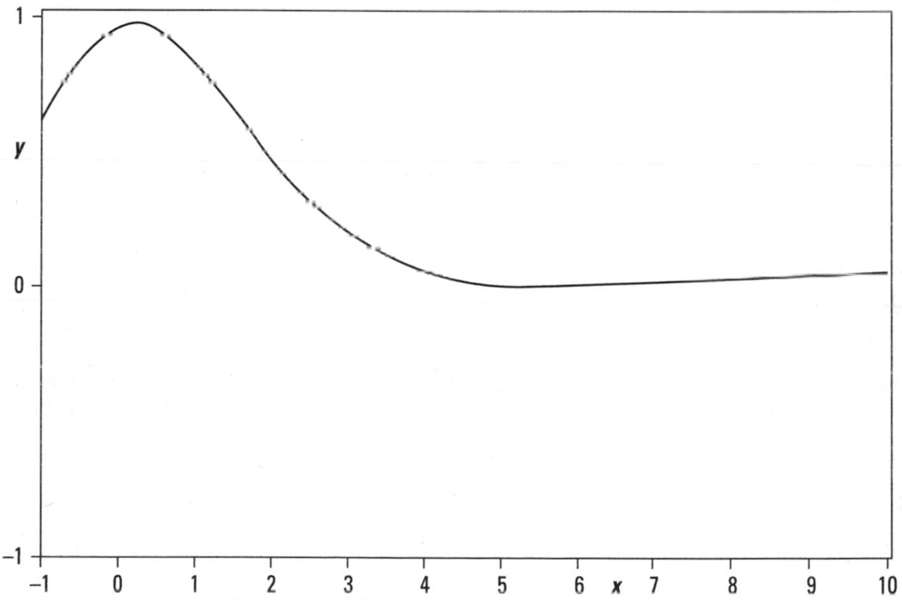


**Figure 5-2:** The solution to an equation with complex roots.

## Identical real roots

In the previous sections, I cover the cases where the characteristic equation has real and distinct roots as well as complex roots. All that's left is the case where the characteristic equation has two real roots that are identical to each other. You get the fundamentals and an example in the following sections.

### The basics

When the roots of a characteristic equation are identical, the discriminant of the quadratic equation, $b^2 - 4ac$, equals zero, which means that:

$$r_1 = -b/2a$$

and

$$r_2 = -b/2a$$

However, now you have a problem. Why? Because second order differential equations are supposed to have two expressions in their solution. Because $r_1 = r_2$, you get this:

$$y_1 = c_1 e^{-bx/2a} \qquad y_2 = c_2 e^{-bx/2a}$$

These solutions differ only by a constant, so they aren't really different at all; you could write either $y_1 + y_2$ or $y_1 - y_2$ as:

$$y = ce^{-bx/2a}$$

So, as you can see, you really only have one solution in this case. How can you find another? Here's the traditional way of solving this issue: So far, you've been finding solutions to second order linear differential equations by assuming the solution is of the following form:

$$y(x) = ce^{rx}$$

However, consider this: What if you replaced the constant with a function of $x$, $m(x)$, instead? Doing so would let you handle more general second order linear differential equations. By doing this, you get:

$$y(x) = m(x)e^{rx}$$

In this case, $r = -b/2a$, so you get this form of the solution:

$$y(x) = m(x)e^{-bx/2a}$$

Differentiating the equation gives you:

$$y'(x) = m'(x)\,e^{-bx/2a} - \frac{b}{2a}\,m(x)\,e^{-bx/2a}$$

You also need $y''(x)$ to substitute into the differential equation. Doing so gives you:

$$y''(x) = m''(x)e^{-bx/2a} - \frac{b}{2a}\,m'(x)e^{-bx/2a} + \frac{b^2}{4a^2}\,m(x)e^{-bx/2a}$$

Substituting $y(x)$, $y'(x)$, and $y''(x)$ into the following differential equation:

$$ay'' + by' + cy = 0$$

gives you (after rearranging the terms):

$$a\,m''(x) + (b-b)\,m'(x) + \left(\frac{b^2}{4a} - \frac{b^2}{2a} + c\right)m(x) = 0$$

So $b - b = 0$ in the second term, along with combining the fractions in the third term, gives you:

$$a\,m''(x) + \left(c - \frac{b^2}{4a}\right)m(x) = 0$$

TIP

Here's the trick: Note that $c - b^2/4a$ is the discriminant of the characteristic equation, and in this case (the case of real identical roots), the discriminant equals zero. So you have:

$$a\,m''(x) = 0$$

or simply, dividing by $a$:

$$m''(x) = 0$$

Wow, that looks pretty easy after all. After integrating, you get this final form for $m(x)$:

$$m(x) = d_1x + d_2$$

where $d_1$ and $d_2$ are constants.

REMEMBER

Because the second solution you've been looking for is $y_2(x) = m(x)e^{rx} = d_1xe^{rx} + d_2e^{rx}$, here's the general solution of the differential equation in the case where the characteristic equation's roots are real and identical:

$$y(x) = c_1xe^{-bx/2a} + c_2e^{-bx/2a}$$

where $c_1$ and $c_2$ are constants.

This trick, it turns out, can often be used to find a second solution to a second order linear homogeneous differential equation if you already know one solution. In fact, trying a solution of the form $y(x) = m(x)f(x)$ is so useful that it has been given a name: the method of *reduction of order.* I discuss this method in more detail later in this chapter.

### An example

Want to see an example so that the identical real root concept can really take hold? Take a look at this differential equation:

$$y'' + 2y' + y = 0$$

where:

$$y(0) = 1$$

and:

$$y'(0) = 1$$

The characteristic equation is:

$$r^2 + 2r + 1 = 0$$

You can factor this equation into:

$$(r + 1)\,(r + 1) = 0$$

So the roots of the characteristic equation are identical, –1 and –1.

Now the solution is of the following form:

$$y(x) = c_1 x e^{-bx/2a} + c_2 e^{-bx/2a}$$

To figure out $c_1$ and $c_2$, use the initial conditions. Substituting into the equation gives you:

$$y(0) = c_2 = 1$$

Differentiating the $y(x)$ equation gives you $y'(x)$; you also can substitute $c_2 = 1$ to get:

$$y'(x) = c_1 e^{-bx/2a} - \frac{bc_1}{2a} x e^{-bx/2a} - \frac{b}{2a} e^{-bx/2a}$$

From the initial conditions, $y'(0)$ equals:

$$y'(0) = c_1 - 1 = 1$$

So $c_1 = 2$, giving you the following general solution:

$$y(x) = 2xe^{-x} + e^{-x}$$

And that's that. You can find a graph of this solution in Figure 5-3.

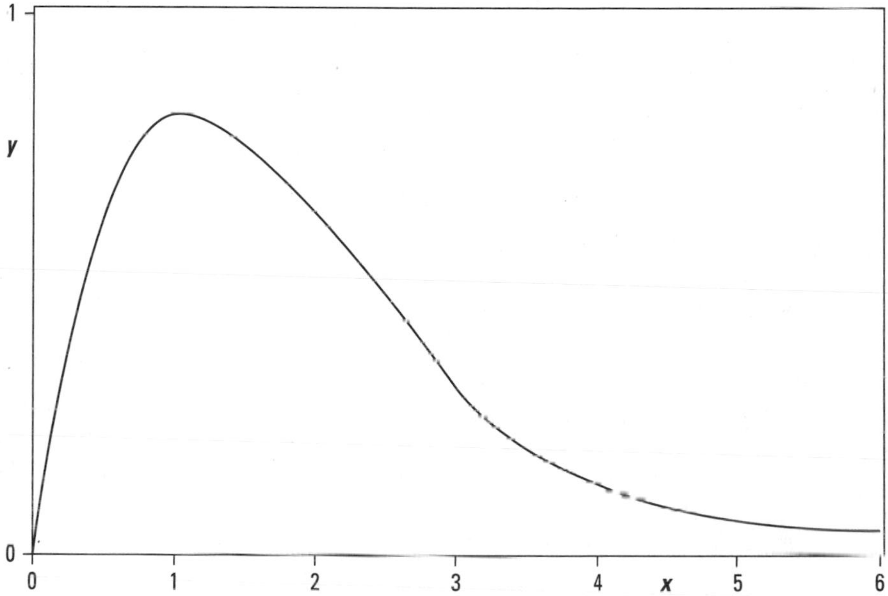


**Figure 5-3:** The solution to an equation with identical real roots.

# Getting a Second Solution by Reduction of Order

If you know one solution to a second order differential equation, you can get a first order differential equation for the second solution — or rather, for the derivative of the second solution, which you can then integrate to get the actual second solution. That's where the term *reduction of order* comes from. The method is very cool, as you find out in the following sections.

## Seeing how reduction of order works

What is the *reduction of order method?* Basically it says that if you already know a solution, $y_1(x)$, to a differential equation of this form:

$$y'' + p(x)y' + q(x) = 0$$

you can substitute the following expression into the equation to search for a second solution:

$$y(x) = m(x)y_1(x)$$

Note that the original equation isn't restricted to differential equations with constant coefficients. However, it is homogeneous (in other words, it equals zero).

Differentiating $y(x) = m(x)y_1(x)$ with respect to $x$ gives you:

$$y'(x) = m'(x)y_1(x) + m(x)y_1'(x)$$

You also need the second derivative of $y(x) = m(x)y_1(x)$, which looks like this:

$$y''(x) = m''(x)y_1(x) + m'(x)y_1'(x) + m'(x)y_1'(x) + m(x)y_1''(x)$$

The middle two terms are the same, so the equation becomes:

$$y''(x) = m''(x)y_1(x) + 2m'(x)y_1'(x) + m(x)y_1''(x)$$

Putting $y'(x)$ and $y''(x)$ into $y'' + p(x)y' + q(x) = 0$ gives you the following equation:

$$y_1m'' + (2y_1' + py_1)m' + (y_1'' + py_1' + qy_1) = 0$$

I bet you recognize the expression in the second set of parentheses here. And guess what? That's the point! That expression is $y_1'' + py_1' + qy_1$, which is just $y_1$ substituted into the original differential equation. But $y_1$ is a solution of that homogeneous differential equation, so the expression in parentheses equals 0!

So you end up with this equation:

$$y_1m'' + (2y_1' + py_1)m' = 0$$

It doesn't look pretty, but examine what you have here: This is actually a first order differential equation in $m'(x)$. If you substitute $n(x) = m'(x)$ into the equation, you get this:

$$y_1 n'(x) + (2y_1' + py_1)n(x) = 0$$

This final equation is a first order differential equation in $n(x)$, and it can often be solved as a first order linear differential equation or as a separable differential equation (see Chapters 2 and 3 for more about these types of equations).

## Trying out an example

In this section, you can see reduction of order at work in an example. Take a look at this second order differential equation:

$$2x^2y'' + xy' - y = 0$$

How do you tackle this one? I walk you through the process in the following sections.

### The first solution

TIP

In the equation $2x^2y'' + xy' - y = 0$, note how each successive derivative $y'$ and $y''$ is multiplied by a new power of $x$. This suggests that differentiating a solution of this differential equation generates a new power of $x$ in the denominator (and that the $x$ and $x^2$ cancel out). So try a first solution of the form:

$$y_1 = 1/x^n$$

Substituting this solution into the equation gives you:

$$2x^2n(n+1)/x^{n+2} - xn/x^{n+1} - 1/x^n = 0$$

By cancelling $1/x^n$, you get:

$$2n(n+1) - n - 1 = 0$$

or:

$$2n^2 - n - 1 = 0$$

The quadratic equation gives you these roots:

$$\frac{1 \pm 3}{4}$$

Taking the root $r = 1$, you see that:

$$y_1 = 1/x$$

is a solution of $2x^2y'' + xy' - y = 0$. (You can take a look at the other root, $r = -½$, in the later section "A little shortcut.")

### *The second solution*

In the previous section, you find the first solution, $y_1 = 1/x$. Now you use the reduction of order method to find a second solution. This means that you're now looking for a solution of the following form:

$$y_2 = m(x)/x$$

The first derivative of this solution is:

$$y_2' = m'(x)/x - m(x)/x^2$$

The second derivative is:

$$y_2'' = m''(x)/x - 2m'(x)/x_2 + 2m(x)/x^3$$

Substituting $y'$ and $y''$ into $2x^2y'' + xy' - y = 0$ gives you (after the algebra settles):

$$2xm'' - m' = 0$$

To make this equation a little easier to work with, set $m'(x) = n(x)$, which gives you this:

$$2x\,\frac{dn}{dx} = n$$

Separating the variables results in the following equation:

$$\frac{dn}{n} = \frac{dx}{2x}$$

or:

$$\ln(n) = \ln(x)/2$$

Exponentiating both sides of the equation gives you:

$$n(x) = kx^{1/2}$$

where $k$ is a constant. Because $n(x) = m'(x)$, you know that:

$$m(x) = cx^{3/2} + d$$

where $c$ and $d$ are constants.

This solves for $m(x)$. Now because, by $y_2 = m(x)/x$:

$$y_2 = m(x)\, y_1 = m(x)/x$$

you know that:

$$y_2 = c_1x^{1/2} + c_2/x$$

Note that the second term here can be absorbed together with $y_1$ (which is $1/x$). So the complete solution is (I've absorbed the constants together as needed):

$$y = y_1 + y_2 = c_1x^{1/2} + c_2/x$$

### A little shortcut

You may have noticed that in the previous example, you didn't actually need to use reduction of order. I'll show you what I mean. The differential equation you were solving looked like this:

$$2x^2y'' + xy' - y = 0$$

You guessed that the solution was of the following form:

$$y = 1/x^n$$

Substituting this into the differential equation gave you:

$$2n^2 - n - 1 = 0$$

which has these roots:

$$\frac{1 \pm 3}{4}$$

So the roots are 1 and –½, which means that you did in fact have two solutions already:

$$y_1 = c_1/x$$

and:

$$y_2 = c_2x^{1/2}$$

# Putting Everything Together with Some Handy Theorems

A couple of formal theorems, which I explain in the following sections, lock down exactly how to find the general solutions to second order linear homogeneous differential equations and formalize the information from earlier sections.

## Superposition

I'll start off with the theorem on superposition of solutions, which formally says this:

**If you have the second order linear homogeneous differential equation:**

$$y'' + p(x)y' + q(x) = 0$$

**and have two solutions, $y_1(x)$ and $y_2(x)$, then any linear combinations of these solutions:**

$$y = c_1y_1 + c_2y_2$$

**where $c_1$ and $c_2$ are constants, is also a solution of the differential equation.**

In plain English, this theorem simply says that if you have two solutions to a second order linear differential equation, a linear combination of those two solutions is also a solution. You actually use this theorem throughout the first part of this chapter as you solve second order linear homogeneous differential equations for two solutions, $y_1$ and $y_2$, and give the general solution as a linear combination of the two.

Here's another example; say that you have the following second order linear differential equation:

$$y'' - y = 0$$

You can guess that the solution is of the form $y = e^{nx}$ and that substituting it in gives you:

$$n^2e^{nx} - e^{nx} = 0$$

or:

$$n^2 - 1 = 0$$

Therefore:

$$n = \pm 1$$

So the two solutions you've worked out are:

$$y_1 = c_1 e^x$$

and

$$y_2 = c_2 e^{-x}$$

According to this theorem, the following linear *superposition* of these solutions is also a solution:

$$y = c_1 e^x + c_2 e^{-x}$$

You can verify your solution by substituting it into the original equation:

$$y'' - y = c_1 e^x + c_2 e^{-x} - (c_1 e^x + c_2 e^{-x}) = 0$$

## Linear independence

Earlier in this chapter, you work problems to find the solutions $y_1$ and $y_2$ to second order linear homogeneous differential equations, but it turns out that there's a restriction on them — they have to be *linearly independent.* What, exactly, does that mean?

If you have two functions, $f(x)$ and $g(x)$, they're linearly dependent in an interval $I$ if for all $x$ in $I$ you can find two constants, $c_1$ and $c_2$, such that this equation is true:

$$c_1 f(x) + c_2 g(x) = 0$$

In other words, $f(x)$ just differs from $g(x)$ by a constant.

If the functions $f$ and $g$ aren't linearly dependent in the interval $I$, they're linearly independent in the interval $I$. In other words, if it's impossible to find two constants such that the previous equation is true for $f(x)$ and $g(x)$ in the interval $I$, then $f(x)$ and $g(x)$ are linearly independent in the interval $I$.

Practically speaking, if one function isn't just a multiple of another function, the two are linearly independent.

Here's the formal theorem:

**If you have the second order linear homogeneous differential equation:**

$$y'' + p(x)y' + q(x) = 0$$

**and have two linearly independent solutions, $y_1(x)$ and $y_2(x)$, then any linear combinations of these solutions:**

$$y = c_1y_1 + c_2y_2$$

**where $c_1$ and $c_2$ are constants, is the general solution of the differential equation. Every solution of the differential equation can be expressed in the form of $y = c_1y_1 + c_2y_2$.**

In other words, you can't just use any two solutions, $y_1$ and $y_2$, to a second order linear homogeneous differential equation. Instead, those two solutions have to be linearly independent so that a linear combination of them is a general solution of the differential equation.

Clear as mud, right? Well, here's an example to wrap your brain around. Say that you have the following differential equation:

$$y'' + 6y' + 8y = 0$$

What's the general solution of this equation? Well, guessing that the solutions are of the form $y = e^{rx}$, you get the following equation by substituting the solution into the differential equation:

$$r^2e^{rx} + 6re^{rx} + 8e^{rx} = 0$$

The characteristic equation is:

$$r^2 + 6r + 8 = 0$$

which can be factored into:

$$(r + 4)\ (r + 2)$$

So the roots of the characteristic equation are –4 and –2, and your solutions are of the form:

$$y_1 = c_1e^{-4x}$$

and

$$y_2 = c_2e^{-2x}$$

These solutions are linearly independent — they don't just differ by a multiplicative constant — so by the theorem of linear independence, the general solution to the differential equation has the following form:

$$y = c_1 e^{-4x} + c_2 e^{-2x}$$

# The Wronskian

You may have heard of the Wronskian, especially if you aspire to be a Differential Equations Wizard (and doesn't everyone?). But if you haven't heard of it, you're probably wondering what it is. Briefly put, *the Wronskian* is the determinant of an array of coefficients that lets you determine the linear independence of solutions to a differential equation.

Okay, that was a mouthful. What does it mean in layman's terms? I explain what you need to know and provide some examples in the following sections.

## Arrays and determinants

Say that you have a second order linear homogeneous differential equation with the following general solution:

$$y = c_1 y_1 + c_2 y_2$$

And assume that you have these initial conditions

$$y(x_0) = y_0$$

and

$$y'(x_0) = y'_0$$

The initial conditions mean that the constants $c_1$ and $c_2$ have to satisfy these equations:

$$y_o = c_1 y_1(x_0) + c_2 y_2(x_0)$$

and:

$$y'_0 = c_1 y'_1(x_0) + c_2 y'_2(x_0)$$

Solving this system of two equations for $c_1$ and $c_2$ gives you:

$$c_1 = \frac{y_0 y'_2(x_0) - y'_0 y_2(x_0)}{y_1(x_0) y'_2(x_0) - y'_1(x_0) y_2(x_0)}$$

and:

$$c_2 = \frac{-y_0 y'_1(x_0) + y'_0 y_1(x_0)}{y_1(x_0) y'_2(x_0) - y'_1(x_0) y_2(x_0)}$$

At some point in time, someone got the bright idea that this system looks like the division of two array determinants. As a refresher on determinants, say you have this array:

$$\begin{vmatrix} a & b \\ c & d \end{vmatrix}$$

In the example, the determinant is:

$$\begin{vmatrix} a & b \\ c & d \end{vmatrix} = ad - cb$$

Using this notation, you can write the expressions for $c_1$ and $c_2$ in determinant form:

$$c_1 = \frac{\begin{vmatrix} y_0 & y_2(x_0) \\ y'_0 & y'_2(x_0) \end{vmatrix}}{\begin{vmatrix} y_1(x_0) & y_2(x_0) \\ y'_1(x_0) & y'_2(x_0) \end{vmatrix}}$$

and:

$$c_2 = \frac{\begin{vmatrix} y_1(x_0) & y_0 \\ y'_1(x_0) & y'_0 \end{vmatrix}}{\begin{vmatrix} y_1(x_0) & y_2(x_0) \\ y'_1(x_0) & y'_2(x_0) \end{vmatrix}}$$

In order for $c_1$ and $c_2$ to make sense in these expressions, the denominator must be nonzero, so:

$$W = \begin{vmatrix} y_1(x_0) & y_2(x_0) \\ y'_1(x_0) & y'_2(x_0) \end{vmatrix} = y_1(x_0) y'_2(x_0) - y'_1(x_0) y_2(x_0) \neq 0$$

Here's the Wronskian (W) of the solutions $y_1$ and $y_2$:

$$W = y_1(x_0)y'_2(x_0) - y_1'(x_0)y_2(x_0)$$

## Who was Wronski?

Jósef Maria Hoëné-Wronski (1778–1853) was a Polish philosopher. He studied and published in many fields, such as philosophy, math, and physics, and he also was an inventor and a lawyer.

Despite deep flashes of insight into math and some other fields, Wronski spent much of his life pursuing ultimately unsuccessful topics, such as perpetual motion machines. Unfortunately, during his lifetime, his work was largely ridiculed and ignored. However, since his death, people have come to unearth significant nuggets of thought from his writings.

TECHNICAL STUFF

You may also see the Wronskian referred to as the *Wronskian determinant.* It's named for a Polish mathematician, Jósef Maria Hoëné-Wronski (and you thought mathematicians had bland names). Check out the nearby sidebar "Who was Wronski?" for more info.

### *The formal theorem*

The Wronskian is a measure of linear independence. Consider this, for example: If you have two solutions, $y_1$ and $y_2$, and their Wronskian is nonzero, then those solutions are linearly independent. And that, my friend, means that every solution to the differential equation is a linear combination of those two solutions.

REMEMBER

And that brings me to a formal theorem:

**If you have the second order linear homogeneous differential equation:**

$$y'' + p(x)y' + q(x) = 0$$

**and have two solutions, $y_1(x)$ and $y_2(x)$, where their Wronskian is nonzero at a point $x_0$, then any linear combinations of these solutions:**

$$y = c_1y_1 + c_2y_2$$

**where $c_1$ and $c_2$ are constants, is the general solution of the differential equation. Every solution of the differential equation can be expressed in the form of $y = c_1y_1 + c_2y_2$.**

Here's the gist of this formal theorem: If the Wronskian of your two solutions is nonzero at a point $x_0$, then a linear combination of those solutions contains every solution of the differential equation. In other words, if you have two solutions, $y_1$ and $y_2$, and their Wronskian is nonzero, then this is the *general solution* of the corresponding differential equation:

$$y = c_1 y_1 + c_2 y_2$$

***Note:*** The $y_1$ and $y_2$ form is what's called a *fundamental solution* to the differential equation.

### Example 1

How about some examples to get a hang of the Wronskian? Earlier in this chapter, you tackled this differential equation:

$$y'' + 2y' + y = 0$$

The characteristic equation is:

$$r^2 + 2r + 1 = 0$$

and you can factor that equation this way:

$$(r + 1)\,(r + 1) = 0$$

which means that the roots of the differential equation are identical, –1 and –1. So the solution is of the following form:

$$y(x) = c_1 x e^{-x} + c_2 e^{-x}$$

Now take a look at the Wronskian, which looks like this:

$$W = \begin{vmatrix} y_1(x_0) & y_2(x_0) \\ y'_1(x_0) & y'_2(x_0) \end{vmatrix} = y_1(x_0)\,y'_2(x_0) - y'_1(x_0)\,y_2(x_0)$$

Because:

$$y_1 = c_1 x e^{-x}$$

and

$$y_2 = c_2 e^{-x}$$

the Wronskian equals:

$$W = y_1(x_0)y'_2(x_0) - y_1'(x_0)y_2(x_0) = -c_1 x e^{-x}\, c_2 e^{-x} - (c_1 e^{-x} - c_1 x e^{-x})c_2 e^{-x}$$

or:

$$W = -c_1 c_2 e^{-2x}$$

And because the Wronskian is always nonzero, you don't even have to substitute $x_0$ in; $y_1$ and $y_2$ form a fundamental solution of the differential equation, as long as $c_1$ and $c_2$ are both nonzero.

### *Example 2*

Here's another example. Say that you have two solutions, $y_1$ and $y_2$, to a second order linear homogeneous differential equation such that:

$$y_1 = e^{r_1 x}$$

and

$$y_2 = e^{r_2 x}$$

Your task now is to show that $y_1$ and $y_2$ form a fundamental solution to the differential equation if $r_1 \neq r_2$.

Time to check the Wronskian:

$$W = \begin{vmatrix} y_1(x_0) & y_2(x_0) \\ y'_1(x_0) & y'_2(x_0) \end{vmatrix} = y_1(x_0)\,y'_2(x_0) - y'_1(x_0)\,y_2(x_0)$$

Because $y_1 = e^{r_1 x}$ and $y_2 = e^{r_2 x}$, the Wronskian equals:

$$W = y_1(x_0)\,y'_2(x_0) - y'_1(x_0)\,y_2(x_0) = e^{r_1 x} r_2 e^{r_2 x} - r_1 e^{r_1 x} e^{r_2 x}$$

or:

$$W = (r_2 - r_1)\,e^{(r_1 + r_2)x}$$

Don't forget: You don't have to substitute $x_0$ into the Wronksian, because it's clearly nonzero everywhere as long as $r_1 \neq r_2$. So $y_1$ and $y_2$ form a fundamental solution set of the differential equation.

Because:

$$y_1 = e^{r_1 x}$$

and

$$y_2 = e^{r_2 x}$$

you can see that what the Wronskian is trying to say is that $y_1$ and $y_2$ are linearly independent. In other words, you can't multiply $y_1$ by a constant so that it equals $y_2$ for all $x$ in an interval, such as $-2 < x < 2$, as long as $r_1 \neq r_2$.

### *Example 3*

Here's one final Wronskian example. Earlier in this chapter, you solve this differential equation:

$$2x^2y'' + xy' - y = 0$$

By guessing that the solution was of the following form:

$$y_1 = 1/x^n$$

you came up with this characteristic equation:

$$2n^2 - n - 1 = 0$$

and found this solution:

$$y = c_1x^{1/2} + c_2/x$$

Is this the general solution? You can find out by bringing the Wronskian to the rescue:

$$W = y_1(x_0)\,y'_2(x_0) - y'_1(x_0)\,y_2(x_0) = -c_1x^{1/2}c_2/x^2 - \frac{c_1}{2}x^{-1/2}c_2/x$$

or:

$$W = -c_1c_2x^{-3/2} - \frac{c_1c_2}{2}x^{-3/2}$$

So:

$$W = \frac{-3}{2}c_1c_2x^{-3/2}$$

If $c_1$ and $c_2 \neq 0$, then because $W \neq 0$ for $x > 0$, you can conclude that:

$$y = c_1x^{1/2} + c_2/x$$

is indeed the general solution.

# Chapter 6

# Studying Second Order Linear Nonhomogeneous Differential Equations

### *In This Chapter*

- Focusing on the facts of linear nonhomogeneous second order equations
- Mapping out the method of undetermined coefficients
- Perusing variation of parameters
- Applying nonhomogeneous equations to physics

Chapter 5 is about second order linear homogeneous differential equations of this form:

$$y'' + p(x)y' + q(x)y = 0$$

This equation is homogeneous because it equals zero (and there's no term that only relies on $x$). In this chapter, however, you tackle differential equations of the *nonhomogeneous* form:

$$y'' + p(x)y' + q(x)y = g(x)$$

where $g(x) \neq 0$.

Exciting, isn't it? To start, I provide you with a theorem that gives you the power you need to solve these kinds of equations. After that, I describe some great techniques for working with these equations. I even throw in a few physics examples to show you how these equations apply to real life.

# The General Solution of Second Order Linear Nonhomogeneous Equations

Luckily for mathematicians everywhere, a quick and easy theorem giving the general solution of a second order linear nonhomogeneous differential equation exists. And this theorem is the one that you'll refer to over and over again in this chapter. In the following sections, I explain the theorem and how you can put it to work.

## Understanding an important theorem

REMEMBER

Without further ado, here's the theorem of the general solution of a second order linear nonhomogeneous differential equation:

**The general solution of the nonhomogeneous differential equation:**

$$y'' + p(x)y' + q(x)y = g(x)$$

**is:**

$$y = c_1y_1(x) + c_2y_2(x) + y_p(x)$$

**where $c_1y_1(x) + c_2y_2(x)$ is the general solution of the corresponding homogeneous differential equation:**

$$y'' + p(x)y' + q(x)y = 0$$

**(for example, $y_1$ and $y_2$ are a fundamental set of solutions to the homogeneous equation) and where $y_p(x)$ is a specific solution to the nonhomogeneous equation.**

This very important theorem basically says that to find the general solution to a nonhomogeneous differential equation, you need the sum of the general solution of the corresponding homogeneous differential equation added to a particular solution of the nonhomogeneous differential equation. A *particular solution* is any solution of the nonhomogeneous differential equation. Quite a mouthful isn't it? Not to worry; I explain how to use this theorem in the next section.

## Putting the theorem to work

You need to follow these steps to find the general solution of a second order linear nonhomogeneous equation:

1. **Find the corresponding homogeneous differential equation by setting $g(x)$ to 0.**
2. **Find the general solution, $y = c_1y_1(x) + c_2y_2(x)$, of the corresponding homogeneous differential equation.**

   This solution is referred to as $y_h$.
3. **Find a single solution to the nonhomogeneous equation.**

   This solution is sometimes referred to as the particular (or specific) solution, $y_p$.
4. **The general solution of the nonhomogeneous differential equation is the sum of $y_h + y_p$.**

To see these steps in action, take a look at this nonhomogeneous differential equation:

$$y'' - y' - 2y = e^{3x}$$

To solve this equation, start by getting the homogenous version of it, like this:

$$y'' - y' - 2y = 0$$

You can assume that the solution is of the form $y = e^{rt}$. So, when you substitute that into the differential equation, you get the following characteristic equation (see Chapter 5 for more details):

$$r^2 - r - 2 = 0$$

You can then factor the characteristic equation this way:

$$(r + 1)(r - 2) = 0$$

Now you can see that the roots, $r_1$ and $r_2$, of the characteristic equation are –1 and 2, giving you:

$$y_1 = e^{-x}$$

and:

$$y_2 = e^{2x}$$

So, the general solution to the homogeneous differential equation is given by:

$$y = c_1e^{-x} + c_2e^{2x}$$

You're halfway there! Now you need a particular solution to the original non-homogeneous differential equation:

$$y'' - y' - 2y = e^{3x}$$

Note that $g(x)$ has the form $e^{3x}$ here, so you can assume that the particular solution has this form:

$$y_p(x) = Ae^{3x}$$

In this case, $A$ is an arbitrary coefficient.

Substituting this form into the nonhomogeneous equation gives you:

$$9Ae^{3x} - 3Ae^{3x} - 2Ae^{3x} = e^{3x}$$

or:

$$9A - 3A - 2A = 1$$

So

$$4A = 1$$

and

$$A = \tfrac{1}{4}$$

The particular solution is:

$$y_p(x) = \frac{e^{3x}}{4}$$

Because the general solution to the nonhomogeneous equation is equal to the sum of the general solution of the corresponding homogeneous differential equation and a particular solution of the nonhomogeneous differential equation, you get this equation as the general solution:

$$y = y_h + y_p$$

or:

$$y = c_1e^{-x} + c_2e^{2x} + \frac{e^{3x}}{4}$$

# Finding Particular Solutions with the Method of Undetermined Coefficients

Are there any techniques that allow you to find particular solutions to nonhomogeneous differential equations? Yes, there are! I'll start with the method of undetermined coefficients, which is actually a method you use earlier in this chapter (you just didn't know the fancy name there).

You start by finding a particular solution to the nonhomogeneous differential equation:

$$y'' + p(x)y' + q(x)y = g(x)$$

REMEMBER

The *method of undetermined coefficients* notes that when you find a candidate solution, $y$, and plug it into the left-hand side of the equation, you end up with $g(x)$. Because $g(x)$ is only a function of $x$, you can often guess the form of $y_p(x)$, up to arbitrary coefficients, and then solve for those coefficients by plugging $y_p(x)$ into the differential equation.

TIP

This method works because you're grappling only with $g(x)$ here, and the form of $g(x)$ can often tell you what a particular solution looks like. For instance, if $g(x)$ is in the form of

- **$e^{rx}$**, try a particular solution of the form $Ae^{rx}$, where $A$ is a constant. Because derivatives of $e^{rx}$ reproduce $e^{rx}$, you have a good chance of finding a particular solution this way.
- **a polynomial of order $n$**, try a polynomial of order $n$. For example, if $g(x) = x^2 + 1$, try a polynomial of the form $Ax^2 + B$.
- **a combination of sines and cosines,** $\sin \alpha x + \cos \beta x$, try a combination of sines and cosines with undetermined coefficients, $A \sin \alpha x + B \cos \beta x$. Then plug into the differential equation and solve for $A$ and $B$.

In the following sections, you take a look at some examples that put to work the method of undetermined coefficients, and then you solve actual differential equations using the theorem I explain earlier in this chapter.

## When $g(x)$ is in the form of $e^{rx}$

You see the following equation in its entirety in the earlier section "Putting the theorem to work," but here's how to figure out the particular solution:

$$y'' - y' - 2y = e^{3x}$$

From the form of the term on the right, $g(x)$, you can guess that a particular solution, $y_p(x)$, has the form:

$$y_p(x) = Ae^{3x}$$

Substituting this into the differential equation gives you:

$$9Ae^{3x} - 3Ae^{3x} - 2Ae^{3x} = e^{3x}$$

Dropping $e^{3x}$ out of each term gives you:

$$9A - 3A - 2A = 1$$

So:

$$4A = 1$$

and

$$A = \frac{1}{4}$$

The particular solution for the differential equation is:

$$y_p(x) = \frac{e^{3x}}{4}$$

## When $g(x)$ is a polynomial of order $n$

The example in this section points out how to deal with a particular solution in the form of a polynomial. Your mission, should you choose to accept it, is to find the general solution of the following nonhomogeneous equation:

$$y'' = 9x^2 + 2x - 1$$

where:

$$y(0) = 1$$

and

$$y'(0) = 3$$

### The general solution to the homogeneous equation

The homogeneous equation is simply:

$$y'' = 0$$

And you can integrate to get this equation:

$$y' = c_1$$

Integrating again gives you the general solution to the homogeneous differential equation, $y_h$:

$$y_h = c_1x + c_2$$

### *The particular and general solutions to the nonhomogeneous equation*

To find the general solution to the nonhomogeneous equation, you need a particular solution, $y_p$. The $g(x)$ term in the original equation is $9x^2 + 2x - 1$, so you can assume that the particular solution has a similar form:

$$y_p = Ax^2 + Bx + C$$

where $A$, $B$, and $C$ are constant coefficients that you have to determine. But here's the issue: The supposed form of $y_p$ has terms in common with $y_h$, the general solution of the homogenous equation:

$$y_h = c_1x + c_2$$

$$y_p = Ax^2 + Bx + C$$

Both of these have an $x$ term and a constant term. When $y_h$ and $y_p$ have terms in common — differing only by a multiplicative constant — that isn't good, because those are really part of the same solution. When you add $y_h$ and $y_p$ together, you get this:

$$y = y_h + y_p = Ax^2 + (c_1 + B)x + (c_2 + C)$$

which can be rewritten:

$$y = y_h + y_p = Ax^2 + cx + d$$

where $c$ and $d$ are constants. As you can see, the terms $Bx + C$ in $y_p$ really don't add anything.

The way to handle this issue is to multiply $y_p$ by successive powers of $x$ until you don't have any terms of the same power as in $y_h$. For example, multiply $y_p$ by $x$ to get:

$$y_p = Ax^3 + Bx^2 + Cx$$

This equation still isn't good enough, however. After all, the $Cx$ term overlaps with the $c_1x$ term in $y_h$. So, you have to multiply by $x$ again to get:

$$y_p = Ax^4 + Bx^3 + Cx^2$$

This solution has no terms in common with the homogeneous general solution, $y_h$, so you're in business.

Substituting $y_p = Ax^4 + Bx^3 + Cx^2$ into $y'' = 9x^2 + 2x - 1$ gives you:

$$12Ax^2 + 6Bx + 2C = 9x^2 + 2x - 1$$

Comparing coefficients of like terms gives you:

$$12A = 9$$

$$6B = 2$$

$$2C = -1$$

which means that:

$$A = 3/4$$

$$B = 1/3$$

$$C = -1/2$$

So the particular solution is:

$$y_p = \frac{3}{4}x^4 + \frac{1}{3}x^3 - \frac{1}{2}x^2$$

Therefore, the general solution is:

$$y = y_h + y_p = c_1x + c_2 + \frac{3}{4}x^4 + \frac{1}{3}x^3 - \frac{1}{2}x^2$$

Or, after rearranging to make things look pretty, you get:

$$y = y_h + y_p = \frac{3}{4}x^4 + \frac{1}{3}x^3 - \frac{1}{2}x^2 + c_1x + c_2$$

where you can get $c_1$ and $c_2$ by using the initial conditions. Substituting $y(0) = 1$ gives:

$$y(0) = 1 = c_2$$

So $c_2 = 1$. Taking the derivative of your general solution gives you:

$$y' = 3x^3 + x^2 - x + c_1$$

And substituting the initial condition $y'(0) = 3$ gives you:

$$y'(0) = 3 = c_1$$

Here's the general solution with all the numbers filled in:

$$y = \frac{3}{4}x^4 + \frac{1}{3}x^3 - \frac{1}{2}x^2 + 3x + 1$$

# When g(x) is a combination of sines and cosines

Combinations of sines and cosines can be a bit tricky, but with the help of the following sections, you can solve equations featuring these combinations in a snap.

### Example 1

Find a particular solution to the following non-homogeneous equation:

$$y'' - y' - 2y = \sin 2x$$

This example looks as though a particular solution may be of the form $y_p = A\sin 2x + B\cos 2x$. There's one way to find out if you're correct; you can substitute this solution into your equation to get:

$$(-6A + 2B)\sin 2x + (-6B - 2A)\cos 2x = \sin 2x$$

You can now equate the coefficients of sin $2x$ and cos $2x$ to get these equations:

$$-6A + 2B = 1$$

and

$$-6B - 2A = 0$$

Multiplying $-6B - 2A = 0$ by $-3$ and adding the result to $6A + 2B = 1$ gives you $20B = 1$. So $B = \frac{1}{20}$. Substituting that number into $-6B - 2A = 0$ gives you the following equation:

$$\frac{-6}{20} - 2A = 0$$

So $A = -3/20$. Therefore, a particular solution of the original nonhomogeneous equation is:

$$y_p = \frac{-3}{20}\sin 2x + \frac{\cos 2x}{20}$$

### *Example 2*

Here's another example using sines and cosines. Try finding a particular solution of the following equation:

$$y'' + 16y = 4\cos 4x$$

Assume that the solution looks like this:

$$y_p = A\cos 4x + B\sin 4x$$

So far, so good. Now, however, when you plug this solution into the differential equation, you get the following result:

$$(16A - 16A)\cos 4x + (16B - 16B)\sin 4x = 4\cos 4x$$

Hang on a minute here! The coefficients of cos $4x$ and sin $4x$ on the left are both zero! So, there are no solutions that are a combination of a sine and a cosine that solve the differential equation. You can see why if you take a look at the corresponding homogeneous differential equation:

$$y'' + 16y = 0$$

The general solution to this homogeneous equation is:

$$y = c_1\cos 4x + c_2\sin 4x$$

TIP

This general solution has the same form as your attempted particular solution: $A\cos 4x + B\sin 4x$. But because that's the homogeneous solution, it isn't going to be a particular solution to the nonhomogeneous equation. Instead, you have to find another version of $y_p$ — one that will result in sin $4x$ and cos $4x$ terms when differentiated. The simplest form (besides $y_p = A\cos 4x + B\sin 4x$) that can do that is:

$$y_p = Ax\cos 4x + Bx\sin 4x$$

Give it a try. Here's what $y'_p$ looks like:

$$y'_p = A\cos 4x - 4Ax\sin 4x + B\sin 4x + 4Bx\cos 4x$$

And $y''_p$ looks like this:

$$y''_p = -4A \sin 4x - 4A \sin 4x - 16Ax \cos 4x + 4B \cos 4x + 4B \cos 4x - 16Bx \sin 4x$$

Substituting $y''_p$ into the original nonhomogeneous differential equation gives you this result:

$$-4A \sin 4x - 4A \sin 4x - 16Ax \cos 4x + 4B \cos 4x + 4B \cos 4x - 16Bx \sin 4x + 16Ax \cos 4x + 16Bx \sin 4x = 4 \cos 4x$$

Wow! Thankfully, this collapses to:

$$-8A \sin 4x + 8B \cos 4x = 4 \cos 4x$$

So, $A = 0$ and $B = ½$, giving you the following particular solution:

$$y_p = \frac{x}{2} \sin 4x$$

And that's that. The general solution to this differential equation is therefore:

$$y = c_1 \cos 4x + c_2 \sin 4x + \frac{x}{2} \sin 4x$$

## When g(x) is a product of two different forms

REMEMBER

Here's a neat trick. If your $g(x)$ term is a product of $e^{rx}$ with sin $x$ and cos $x$ or a polynomial, you should attempt a particular solution that's a similar product, using coefficients whose values have yet to be determined.

Ready to see this handy trick at work? Take a look at this differential equation.

$$4y'' + y = 5e^x \cos 2x$$

Assume that a particular solution is the product of $e^x$ and cos $2x$ and sin $2x$:

$$y_p = Ae^x \cos 2x + Be^x \sin 2x$$

Here's the derivative $y'_p$:

$$y'_p = Ae^x \cos 2x - 2Ae^x \sin 2x + Be^x \sin 2x + 2Be^x \cos 2x$$

And here's the second derivative $y''_p$:

$$y''_p = Ae^x \cos 2x - 2Ae^x \sin 2x - 2Ae^x \sin 2x - 4Ae^x \cos 2x + Be^x \sin 2x + 2Be^x \cos 2x + 2Be^x \cos 2x - 4Be^x \sin 2x$$

or:

$$y''_p = -3A\, e^x \cos 2x - 4Ae^x \sin 2x - 3Be^x \sin 2x + 4Be^x \cos 2x = e^x \cos 2x\,(-3A + 4B) + e^x \sin 2x\,(-4A - 3B)$$

Substituting this into your original differential equation gives you the following:

$$4(-3Ae^x \cos 2x - 4Ae^x \sin 2x - 3Be^x \sin 2x + 4Be^x \cos 2x) + (Ae^x \cos 2x + Be^x \sin 2x) = 5e^x \cos 2x$$

or:

$$-12Ae^x \cos 2x - 16Ae^x \sin 2x - 12Be^x \sin 2x + 16Be^x \cos 2x + Ae^x \cos 2x + Be^x \sin 2x = 5e^x \cos 2x$$

Collecting terms gives you this equation:

$$(-11A + 16B)e^x \cos 2x + (-16A - 11B)e^x \sin 2x = 5e^x \cos 2x$$

So, matching the coefficients of cos $2x$ and sin $2x$ gives you these equations:

$$-11A + 16B = 5$$

and

$$-16A - 11B = 0$$

From this equation, $A = -11B/16$, and by substituting it into the previous equation you get this:

$$-121B + 256B = 80$$

So, $135B = 80$, or $B = 16/27$. Figuring out $A$ yields:

$$-16A - 176/27 = 0$$

And $A = -11/27$. Therefore the particular solution, $y_p$, is:

$$y_p = \frac{-11}{27} e^x \cos 2x + \frac{16}{27} e^x \sin 2x$$

# Breaking Down Equations with the Variation of Parameters Method

What if $g(x)$ isn't in one of the forms that I discuss in the previous sections? What if you can't get the method of undetermined coefficients to work? In either case, you can always turn to the *variation of parameters method.*

So what's this variation of parameters technique? It's a clever one! Say that you have the following differential equation:

$$y'' + p(x)y' + q(x)y = g(x)$$

Now assume that you know the solution to the corresponding homogeneous equation:

$$y'' + p(x)y' + q(x)y = 0$$

The general homogenous solution is:

$$y_h = c_1y_1 + c_2y_2$$

Here's the first trick: You now replace the constants $c_1$ and $c_2$ with functions $u_1(x)$ and $u_2(x)$ to find a particular solution. This is what the equation looks like:

$$y_p = u_1(x)y_1 + u_2(x)y_2$$

Then you try to find the functions $u_1(x)$ and $u_2(x)$ such that this equation is a particular solution of the nonhomogeneous differential equation (not the homogeneous differential equation).

Here's the second trick: Substituting this equation into your original nonhomogeneous equation is going to give you one equation in two unknowns, $u_1(x)$ and $u_2(x)$, as well as their first two derivatives. You need a second constraint on the values of $u_1(x)$ and $u_2(x)$ as well. And because you're looking only for a single particular solution, you can choose that constraint on $u_1(x)$ and $u_2(x)$ to make the math easier.

In the following sections, I explain the basics of the variation of parameters method and take you step by step through some interesting examples. I also discuss the relationship of this method to one called the Wronskian (see Chapter 5 for details).

## Nailing down the basics of the method

First up: You need $y'$ and $y''$ to substitute into your original nonhomogeneous equation. Start with $y'$, the first derivative of $y_p = u_1(x)y_1 + u_2(x)y_2$:

$$y' = u'_1(x)y_1 + u_1(x)y'_1 + u'_2(x)y_2 + u_2(x)y'_2$$

TIP

Here's where the second trick comes in (the one designed to make the math easier). Because you get to choose the second constraint on $u_1(x)$ and $u_2(x)$, choose the constraint so that:

$$u'_1(x)y_1 + u'_2(x)y_2 = 0$$

The second trick makes the math easier because now the first derivative becomes:

$$y' = u_1(x)y'_1 + u_2(x)y'_2$$

Quite a bit simpler, right? Now for $y''$:

$$y'' = u'_1(x)y'_1 + u_1(x)y''_1 + u'_2(x)y'_2 + u_2(x)y''_2$$

Now substitute $y$, $y'$, and $y''$ into the nonhomogeneous equation. The resulting equation is going to look messy, but don't fret because something good happens:

$$u_1(x)[y_1'' + p(x)y_1' + q(x)y_1] + u_2(x)[y_2'' + p(x)y_2' + q(x)y_2] + u'_1(x)y_1' + u'_2(x)y_2' = g(x)$$

Note that the first two terms are zero (that's the good thing that happens), because $y_1$ and $y_2$ are solutions of the homogeneous equation. This means that you're left with:

$$u'_1(x)y_1' + u'_2(x)y_2' = g(x)$$

You now have two equations in $u'_1(x)$ and $u'_2(x)$:

$$u'_1(x)y_1(x) + u'_2(x)y_2(x) = 0$$

$$u'_1(x)y_1'(x) + u'_2(x)y_2'(x) = g(x)$$

REMEMBER

Using these equations, you can solve for $u'_1(x)$ and $u'_2(x)$. Then you can integrate them, and you'll have a particular solution to the original nonhomogeneous differential equation, because where $y_1$ and $y_2$ are linearly independent solutions of the homogeneous equation:

$$y_p = u_1(x)y_1(x) + u_2(x)y_2(x)$$

## Solving a typical example

To help solidify the basics in your mind, put them to work with the following differential equation:

$$y'' + 4y = \sin^2 2x$$

The homogenous equation is:

$$y'' + 4y = 0$$

The general solution to the homogeneous equation is the following (I'll skip the details here; you can find out how to complete this step earlier in this chapter):

$$y_h = c_1 \cos 2x + c_2 \sin 2x$$

So:

$$y_1 = \cos 2x$$

and

$$y_2 = \sin 2x$$

This means that you'll be searching for a particular solution of the following form:

$$y_p = u_1(x) \cos 2x + u_2(x) \sin 2x$$

where you need to determine $u_1(x)$ and $u_2(x)$. The method of variation of parameters tells you that:

$$u'_1(x)y_1(x) + u'_2(x)y_2(x) = 0$$

$$u'_1(x)y_1'(x) + u'_2(x)y_2'(x) = g(x)$$

Substituting in for $y_1$ and $y_2$ gives you these equations:

$$u'_1(x) \cos 2x + u'_2(x) \sin 2x = 0$$

$$-2u'_1(x) \sin 2x + 2u'_2(x) \cos 2x = \sin^2 2x$$

Solving these equations for $u'_1(x)$ and $u'_2(x)$ gives you:

$$u'_1(x) = \frac{-\sin^3 2x}{2}$$

and:

$$u'_2(x) = \frac{\sin^2 2x \cos 2x}{2}$$

You can integrate to find $u_1(x)$ and $u_2(x)$:

$$u_1(x) = \frac{\cos 2x}{4} - \frac{\cos^3 2x}{12}$$

and:

$$u_2(x) = \frac{\sin^3 2x}{12}$$

Why aren't you using any constants of integration here? Because you need only one particular solution, so you can choose the constants of integration to equal zero, which means that the particular solution is (after the algebra and trig dust settles):

$$y_p(x) = \frac{\cos^2 2x}{6} + \frac{\sin^2 2x}{12}$$

So, the general solution of your original nonhomogeneous equation is:

$$y = y_h + y_p$$

which means it looks like this:

$$y = c_1 \cos 2x + c_2 \sin 2x + \frac{\cos^2 2x}{6} + \frac{\sin^2 2x}{12}$$

Beautiful.

## Applying the method to any linear equation

As you can see from the previous sections, the method of variation of parameters can be useful. It's broader in application than the method of undetermined coefficients that I discuss earlier in this chapter. Why? The method of undetermined coefficients works only for a few forms of $g(x)$.

On the other hand, you can use the method of variation of parameters for all linear differential equations (linear in $y$). For second order differential equations like the ones in this chapter, you get a system of two equations to solve; for a system of three differential equations, you get three equations to solve; and so on. The problem comes in the integration of $u'_1(x)$, $u'_2(x)$, and so on, because the integration may not be possible.

Take a look at an example that brings this issue to light. Here's a whopper of a differential equation:

$$\frac{d^2y}{dx^2} - \frac{2}{x}\frac{dy}{dx} + \frac{2y}{x^2} = \frac{1}{x}\ln(x)$$

Why is this equation such a whopper? Well, for starters, it isn't separable (see Chapter 3 for an explanation of separable equations). And to top it off, you can't use the method of undetermined coefficients. But it's linear, so you can use the method of variation of parameters, as you find out in the following sections.

### The general solution of the homogeneous equation

First, take a look at the homogeneous equation:

$$\frac{d^2y}{dx^2} - \frac{2}{x}\frac{dy}{dx} + \frac{2y}{x^2} = 0$$

TIP

You may notice that this equation looks a little more manageable than the original one. After looking at the form of this differential equation and noting that each successive term has another power of $x$ in the denominator, you would likely decide to try a solution of the form $y = x^n$. And by substituting into the differential equation, you get:

$$n(n-1)x^{n-2} - 2nx^{n-2} + 2x^{n-2} = 0$$

After dividing by $x^{n-2}$ and doing a little algebra, you get:

$$n^2 - n - 2n + 2 = 0$$

So:

$$n^2 - 3n + 2 = 0$$

You can solve this with the quadratic equation to get:

$$n = \frac{3 \pm 1}{2}$$

So $n_1 = 1$ and $n_2 = 2$, which means that you have two linearly independent solutions of the homogeneous differential equation:

$$y_1 = x$$

and

$$y_2 = x^2$$

You get the following as the general solution of the homogeneous differential equation:

$$y_h = c_1 x + c_2 x^2$$

### The particular and general solutions of the nonhomogeneous equation

Now you have to find the particular solution, $y_p$, of the original nonhomogeneous differential equation. Why? Because its general solution is the sum of the general solution of the homogeneous differential equation and the particular solution of the nonhomogeneous differential equation:

$$y = y_h + y_p$$

Say you have a linear differential equation of the following form:

$$y'' + p(x)y' + q(x)y = g(x)$$

In this case, it's a doozy of an equation, and:

$$p(x) = \frac{2}{x}$$

$$q(x) = \frac{2}{x^2}$$

$$g(x) = \frac{1}{x}\ln(x)$$

Okay, so you can't use the method of undetermined coefficients. Never fear; this is where the method of variation of parameters comes in. Accordingly, you plan a solution of the following form:

$$y_p = u_1(x)x + u_2(x)x^2$$

The method of variation of parameters gives you:

$$u'_1(x)y_1(x) + u'_2(x)y_2(x) = 0$$

$$u'_1(x)y_1'(x) + u'_2(x)y_2'(x) = g(x)$$

So:

$$u'_1(x)x + u'_2(x)x^2 = 0$$

and:

$$u'_1(x) + u'_2(x)\,2x = \frac{1}{x}\ln(x)$$

From:

$$u'_1(x)x + u'_2(x)x^2 = 0$$

you get:

$$u'_1(x) = -u'_2(x)x$$

And substituting that into the second equation gives you:

$$-u'_2(x)\,x + u'_2(x)\,2x = \frac{1}{x}\ln(x)$$

So with a little combining, you get:

$$u'_2(x) = \frac{1}{x^2}\ln(x)$$

Substituting that equation into $u'_1(x)x + u'_2(x)x^2 = 0$ gives you the following:

$$u'_1(x) = -\frac{1}{x}\ln(x)$$

Now you have $u'_1(x)$ and $u'_2(x)$, and you have to integrate them. (***Tip:*** You can find the integrals in most standard calculus books.) Here are the answers (as in the previous section, neglecting any constants of integration, because you need only one particular solution):

$$u_1(x) = -\frac{1}{2}\ln^2(x)$$

and:

$$u_2(x) = -\frac{1}{x}\ln(x) \quad \frac{1}{x}$$

Substituting that into the following equation:

$$y_p = u_1(x)x + u_2(x)x^2$$

gives you:

$$y_p = -\frac{x}{2}\ln^2(x) - x\ln(x) - x$$

Alright. You're almost there! Because:

$$y = y_h + y_p$$

you know that:

$$y = c_1 x + c_2 x^2 - \frac{x}{2}\ln^2(x) - x\ln(x) - x$$

In fact, you can absorb the final $-x$ term into $c_1x$, giving you the general solution:

$$y = c_1 x + c_2 x^2 - \frac{x}{2}\ln^2(x) - x\ln(x)$$

## What a pair! The variation of parameters method meets the Wronskian

As noted in the previous sections, the method of variation of parameters allows you to tackle linear differential equations, such as this second order differential equation:

$$y'' + p(x)y' + q(x)y = g(x)$$

The method of variation of parameters relies on the solution to the homogeneous equation:

$$y'' + p(x)y' + q(x)y = 0$$

The solution to the homogeneous equation is:

$$y_h = c_1y_1(x) + c_2y_2(x)$$

The method of variation of parameters says that you then try to find a particular solution of the following form:

$$y_p = u_1(x)y_1(x) + u_2(x)y_2(x)$$

Substituting $y_p$ into the differential equation gives you these two equations:

$$u'_1(x)y_1(x) + u'_2(x)y_2(x) = 0$$

$$u'_1(x)y_1{}'(x) + u'_2(x)y_2{}'(x) = g(x)$$

You can formally solve these equations for $u'_1(x)$ and $u'_2(x)$ as follows:

$$u'_1(x) = \frac{-y_2(x)g(x)}{y_1(x)y'_2(x) - y'_1(x)y_2(x)}$$

and:

$$u'_2(x) = \frac{y_1(x)g(x)}{y_1(x)y'_2(x) - y'_1(x)y_2(x)}$$

In fact, it turns out that the denominator is the Wronskian (introduced in Chapter 5), $W$, for $y_1$, $y_2$, and $x$, $W(y_1, y_2)(x)$:

$$W = y_1(x)y'_2(x) - y_1'(x)y_2(x)$$

So, you can write the equations for $u'_1(x)$ and $u'_2(x)$ like this instead:

$$u'_1(x) = \frac{-y_2(x)g(x)}{W(y_1,y_2)(x)}$$

and:

$$u'_2(x) = \frac{y_1(x)g(x)}{W(y_1,y_2)(x)}$$

TIP

Note that dividing by the Wronskian is okay because $y_1$ and $y_2$ are a set of linearly independent solutions, so their Wronskian is nonzero. This means that you can solve for $u_1(x)$ (at least theoretically) like this:

$$u_1(x) = -\int \frac{y_2(x)g(x)}{W(y_1,y_2)(x)}\,dx + c_1$$

And you can solve for $u_2(x)$ like this:

$$u_2(x) = \int \frac{y_1(x)g(x)}{W(y_1,y_2)(x)}\,dx + c_2$$

REMEMBER

Of course, there's no guarantee that you can perform the integrals in these equations. But if you can, you can get $u_1(x)$ and $u_2(x)$, such that a particular solution to the differential equation is:

$$y_p = u_1(x)y_1(x) + u_2(x)y_2(x)$$

The general solution is:

$$y = c_1y_1(x) + c_2y_2(x) + u_1(x)y_1(x) + u_2(x)y_2(x)$$

# Bouncing Around with Springs 'n' Things

Second order differential equations play a big part in elementary physics. They're used in describing the motion of springs and pendulums, electromagnetic waves, heat conduction, electric circuits that contain capacitors and inductors, and so on. I provide a couple of examples of second order differential equations in the following sections.

## A mass without friction

Here I show you the differential equation describing the motion of a mass on the end of a spring. Say, for example, that you have the situation shown in Figure 6-1, where a mass is moving around (without friction) on the end of a spring. In the following sections, I show you how to solve this physics problem with the help of a nonhomogeneous equation.

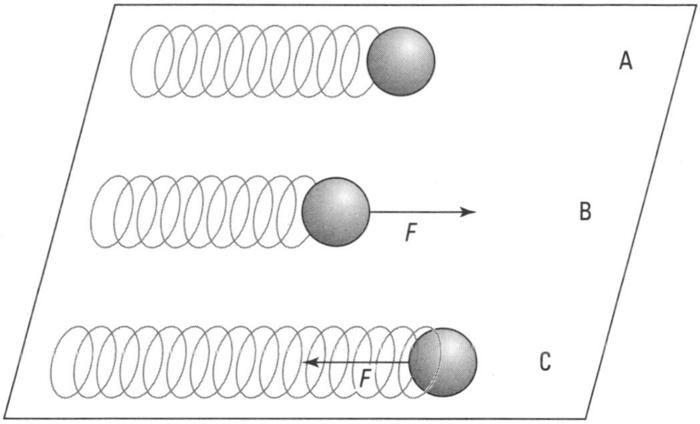


**Figure 6-1:** A spring with a mass moving without friction.

### *Turning the physics into a differential equation*

The force that the spring in Figure 6-1 exerts on the mass is proportional to the amount that the spring is stretched, and the constant of proportionality is called the spring constant, $k$. Thus the force that the spring exerts on the mass is:

$$F = -ky$$

where $k$ is the spring constant (which you have to measure for every spring you want to use) and $y$ is the distance away from the equilibrium position (where the spring is unstretched).

REMEMBER

The minus sign is included to indicate that the force is a *restorative* force, meaning that it always pulls toward the equilibrium position. So the force is always in the opposite of the direction you're pulling. If $x$, the distance from the equilibrium position, is positive, the force exerted by the spring is negative, pulling back toward the equilibrium position.

The force exerted by the spring is proportional to the length by which you're pulling the spring. As you may know from elementary physics, the force on an object is equal to its mass multiplied by its acceleration:

$$F = ma$$

where $m$ is the object's mass, and $a$ is its acceleration.

Here's where the differential equation comes in, because acceleration is the second order derivative of distance with respect to time (symbolized by $t$):

$$a = \frac{d^2y}{dt^2}$$

You can write the equation for force like this:

$$F = m\frac{d^2y}{dt^2}$$

In keeping with the notation I use throughout this book, I'll write the previous equation like this, where the second derivative of distance with respect to time is given by $y''$:

$$F = my''$$

Note that in physics, differentiating by time is often given by putting a dot above the variable being differentiated, and differentiating twice with respect to time is given by placing two dots above the variable being differentiated, like this: ë.

The mass is accelerated by the spring, so $F = ma$ is equal to $F = -ky$. So you have this:

$$my'' = -ky$$

You're back in differential equation territory again, so now you'll take over from the physicists. Here's what this equation looks like in a form you're more used to:

$$my'' + ky = 0$$

This equation is okay, but it's a homogeneous differential equation. And this is, after all, a chapter on nonhomogeneous differential equations. To solve this dilemma, you can add a periodic force, acting on the mass, say $F_0\cos\omega_0 t$. Adding this force makes this a nonhomogeneous differential equation:

$$my'' + ky = F_0 \cos \omega_0 t$$

You're *driving* the mass with this new force, $F_0 \cos \omega_0 t$. This is a periodic force, with period (that is, the time it takes to complete a cycle):

$$T = \frac{2\pi}{\omega_0}$$

What's going to happen now that the mass is subject to the spring force and the new driving force? You can find out by solving $my'' + ky = F_0 \cos \omega_0 t$.

### The general solution to the homogeneous equation

To solve $my'' + ky = F_0 \cos \omega_0 t$, you need to take a look at the corresponding homogeneous equation:

$$my'' + ky = 0$$

Put this equation into standard form, like so:

$$y'' + \frac{ky}{m} = 0$$

In other words, the equation looks like this:

$$y'' = -\frac{ky}{m}$$

Now you need something that changes sign upon being differentiated twice, which means that you need to use sines and cosines. So assume that:

$$y_1 = \cos \omega x$$

and

$$y_2 = \sin \omega x$$

Plugging $y_1$ and $y_2$ into the $y''$ equation, you get:

$$\omega^2 = \frac{k}{m}$$

or:

$$\omega = \sqrt{\frac{k}{m}}$$

So the general homogeneous solution is:

$$y_h = c_1 \cos \omega x + c_2 \sin \omega x$$

### The particular and general solutions to the nonhomogeneous equation

Now you have to find a particular solution to the nonhomogeneous equation:

$$my'' + ky = F_0 \cos \omega_0 t$$

which you can write as:

$$y'' + \frac{ky}{m} = \frac{F_0}{m} \cos \omega_0 t$$

Using the method of undetermined coefficients that I describe earlier in this chapter, you can guess that the particular solution is of the following form:

$$y_p = A \cos \omega_0 t$$

where $A$ is yet to be determined.

**TIP**

You can dispense with the $B \sin \omega_0 t$ term because the differential equation involves only $y''$ and $y$, and $g(x)$ is a cosine. So the $B$ in $B \sin \omega_0 t$ would have to be zero.

Plugging your attempted solution into the nonhomogeneous equation yields:

$$-A\ \omega_0^2 \cos \omega_0 t + \frac{k}{m} A \cos \omega_0 t = \frac{F_0}{m} \cos \omega_0 t$$

or, since $k/m = \omega^2$:

$$-A\ \omega_0^2 \cos \omega_0 t + A\ \omega^2 \cos \omega_0 t = \frac{F_0}{m} \cos \omega_0 t$$

Dividing by $\cos \omega_0 t$ to simplify gives you:

$$-A\ \omega_0^2 + A\ \omega^2 = \frac{F_0}{m}$$

Or:

$$A\left(\omega^2 - \omega_0^2\right) = \frac{F_0}{m}$$

So:

$$A = \frac{F_0}{m\left(\omega^2 - w_0^2\right)}$$

And:

$$y_p = \frac{F_0 \cos \omega_0 t}{m\left(\omega^2 - \omega_0^2\right)}$$

The general solution to the forced spring differential equation is:

$$y = c_1 \cos \omega x + c_2 \sin \omega x + \frac{F_0 \cos \omega_0 t}{m\left(\omega^2 - \omega_0^2\right)}$$

You can see the graph of a representative solution, with $c_1 = c_2 = 1$, $\omega = 1$, $\omega_0 = ½$, and $F_0/m(\omega^2 - \omega_0^2) = 1$, in Figure 6-2.

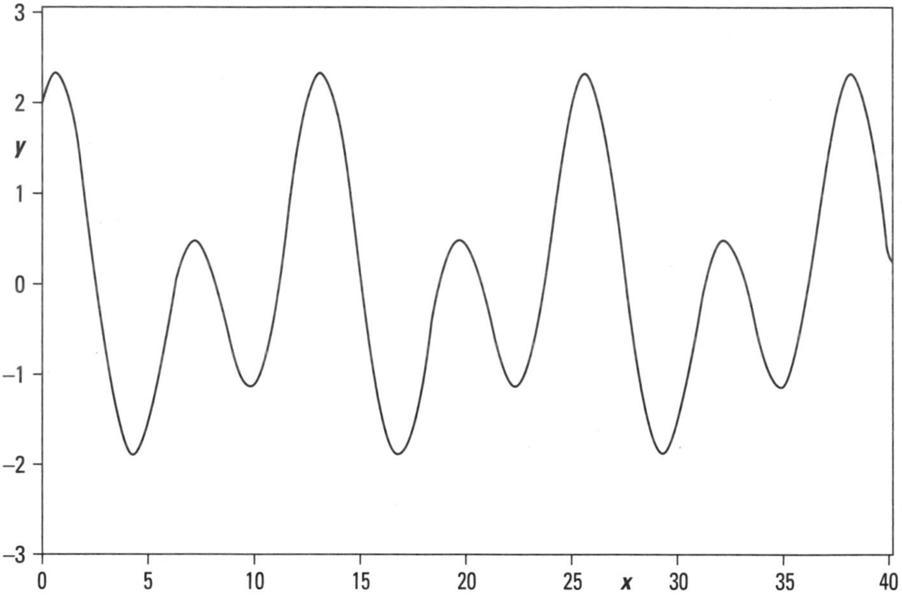


**Figure 6-2:** Graphing the math behind a mass without friction.

## A mass with drag force

In this section is another example, but this time you include a drag force on the mass. (You don't have to read this example if you don't want to. If you do, be forewarned that the algebra gets a little involved.)

A drag force acting on a mass is referred to as *damping*. For example, the mass may be moving through water or heavy fluid, and the damping force tends to slow it down. The damping force is usually proportional to the speed of the mass (the faster it goes, the more damping force there is), and the constant of proportionality is $\gamma$, the damping coefficient. So here's what the differential equation of motion looks like with damping:

$$my'' + \gamma y' + ky = 0$$

You now have a term in $y'$, the speed of the mass on a spring. So how do you solve this one? Well, you can try a solution of the form $y = e^{rt}$ (which includes sines and cosines if $r$ is complex). Plugging this solution for $y$ into the previous equation gives you:

$$mr^2e^{rt} + \gamma re^{rt} + ke^{rt} = 0$$

Canceling out $e^{rt}$ gives you this characteristic equation:

$$mr^2 + \gamma r + k = 0$$

The roots look like this:

$$r_1, r_2 = \frac{-\gamma \pm \left(\gamma^2 - 4mk\right)^{1/2}}{2m}$$

Look at the case where the discriminant, $\gamma^2 - 4mk$, is less than zero. In that case:

$$r_1, r_2 = \frac{\gamma \pm i\left(4mk - \gamma^2\right)^{1/2}}{2m}$$

So substituting $r_1$ and $r_2$ into $e^{rt}$, the solution can be written as:

$$y = e^{-\gamma t/2m}(A \cos \omega t + B \sin \omega t)$$

where:

$$\omega = \frac{\left(4km - \gamma^2\right)^{1/2}}{2m}$$

The solution to the example is an interesting result. It indicates that the motion of the mass is *sinusoidal* (like a sine wave), but it's also multiplied by an exponential that decays with time. So in this case, the mass oscillates with a diminishing amplitude, tending toward zero motion. The usual way that the solution is written is to let $A = C \cos \delta$ and $B = C \sin \delta$, where $\delta$ is called the *phase angle*. Now you can write the solution this way:

$$y = Ce^{-\gamma t/2m} \cos (\omega t - \delta)$$

Writing the solution this way gives you the result in an even neater form. As you can see, in this case, the solution has a decaying sinusoidal form — the $\cos (\omega t - \delta)$ is multiplied by $e^{-\gamma t/2m}$, which goes to zero in time.

This reaction is just what you'd expect from a damped mass on a spring — it would start oscillating, as you'd expect, and then slowly its motion would die away. Think of a mass on a spring immersed in oil, for example.

You can see a representative graph showing the mass's motion in time in Figure 6-3.

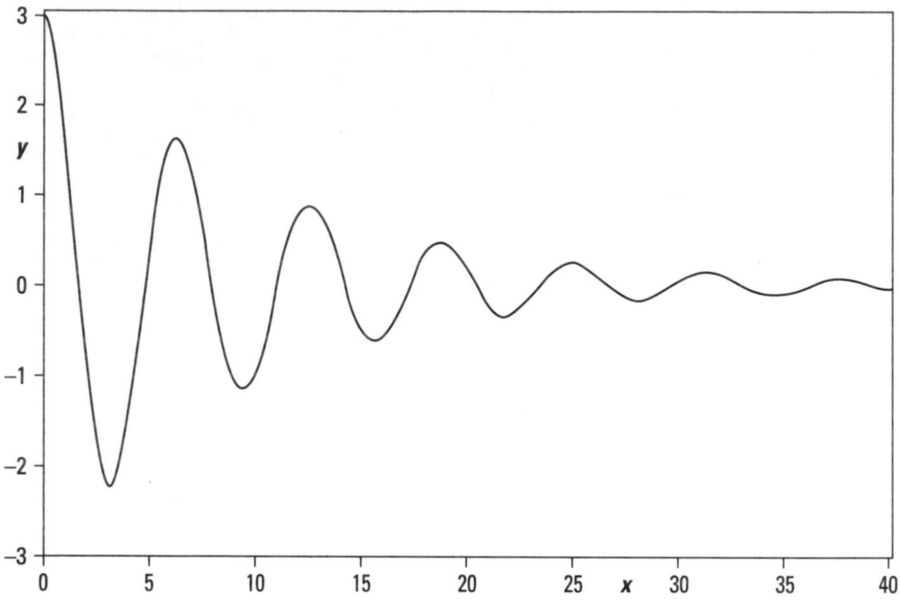


**Figure 6-3:** Graphing the math behind a mass with drag force.

# Chapter 7

# Handling Higher Order Linear Homogeneous Differential Equations

### *In This Chapter*

- Discovering the notation of higher order differential equations
- Introducing the fundamentals of higher order linear homogeneous equations
- Working with real and distinct roots
- Checking out complex roots
- Dealing with duplicate roots

Some technicians from the local nuclear power plant storm into your richly appointed office and say, "We need you to solve a differential equation, and quick."

"Sure," you say, "just let me finish lunch."

"No," they say, "we need the solution *now*." You notice that they keep looking over their shoulders nervously.

"And if you don't get your solution now?" you ask, annoyed.

"Boom," they say.

You put down your sandwich; this might be a rush job after all. "Okay," you say with a sigh, "let me see the differential equation."

They put a piece of paper in front of you, and then you take a look:

$$y''' - 6y'' + 11y' - 6y = 0$$

"It's a *third* order differential equation," they say.

"I can see that," you tell them.

"We only know how to solve up to second order," they wail.

"Many first and second order methods are applicable to higher order differential equations," you say, getting out your clipboard.

"Can you speed things up?" they ask, fidgeting.

"No problem," you say. "The solution looks like this:"

$$y = c_1e^x + c_2e^{2x} + c_3e^{3x}$$

"Wow," they say. "That was quick work." Then the techs start hurrying out the door without paying you so much as a penny. Good thing you love working differential equations.

This chapter introduces higher order (also called *n*th order) differential equations, and by higher order, I mean any order higher than two. With the help of this chapter, you can solve higher order differential equations in a snap, even if they aren't a matter of life and death! I focus on tips and tricks for solving different kinds of higher order linear homogeneous equations (in other words, those that equal zero).

## The Write Stuff: The Notation of Higher Order Differential Equations

Before I get into the nitty-gritty of higher order differential equations in the following sections, you need a primer on their notation. Why? Well, they're slightly different from first and second order equations. For instance, here's an example of a higher order differential equation:

$$\frac{d^4y}{dx^4} - y = 0$$

As you can see, this involves the fourth derivative of $y$ with respect to $x$, so it's a fourth order differential equation. If you've read the first few chapters of this book, you may expect that you can also write the equation like this:

$$y'''' - y = 0$$

Surprise! Instead, this equation is usually written as:

$$y^{(4)} - y = 0$$

Note the terminology: Instead of writing $y''''$, you write $y^{(4)}$. Derivatives up to third order, $y'''$, are commonly written using primes, but when it comes to fourth order and up, go with the $y^{(4)}$ notation. After all, what would you rather see:

$$y^{(9)} - y = 0$$

or:

$$y''''''''' - y = 0$$

In some books, you may see higher order derivatives given with Roman numerals, like this:

$$y^{ix} - y = 0$$

But because Roman numerals are on their way out in common usage today, I stick with the $y^{(9)}$ version here.

# Introducing the Basics of Higher Order Linear Homogeneous Equations

Are you ready? It's time to dig into higher order linear homogeneous differential equations. I start from the beginning in the following sections, with information on their format, solutions, and initial conditions. I also provide some handy theorems to help you on your way.

## The format, solutions, and initial conditions

A general higher order linear differential equation looks like this:

$$\frac{d^n y}{dx^n} + p_1(x)\frac{d^{n-1}y}{dx^{n-1}} + p_2(x)\frac{d^{n-2}y}{dx^{n-2}} + \ldots + p_{n-1}(x)\frac{dy}{dx} + p_n(x)\,y = g(x)$$

This differential equation is linear in all derivatives of $y$ with respect to $x$ (it involves only terms to the power 1), and it's nonhomogeneous because it equals the function $g(x)$.

The homogeneous version of this differential equation looks like this:

$$\frac{d^n y}{dx^n} + p_1(x)\frac{d^{n-1} y}{dx^{n-1}} + p_2(x)\frac{d^{n-2} y}{dx^{n-2}} + \ldots + p_{n-1}(x)\frac{dy}{dx} + p_n(x)\,y = 0$$

where $g(x) = 0$.

Because homogeneous equations equal zero, working with them is a nice way to ease into the world of higher order equations. You take a look at the higher order homogeneous differential equation in this chapter, and then you delve in to the nonhomogeneous version in Chapter 8.

REMEMBER

Say, for example, that you have some solutions of the homogeneous equation — $y_1$, $y_2$, and so on. If $y_1$, $y_2$, and so on are all solutions, then a linear combination of them is also a solution. For instance:

$$y = c_1 y_1 + c_2 y_2 + \ldots c_{n-1} y_{n-1} + c_n y_n$$

In this solution, $c$ stands for various constants. (See the later section "The general solution of a higher order linear homogeneous equation" for more information.)

In addition, higher order differential equations can have initial conditions, just as first order and second order differential equations can. For a higher order differential equation, you can have this many initial conditions:

$$y(x_0) = y_0,\ y'(x_0) = y'_0 \ldots y^{(n-1)}(x_0) = y^{(n-1)}{}_0$$

TIP

Solving differential equations of higher order where $n = 3$ or more is a lot like solving differential equations of first or second order, except that you need more integrations and have to solve larger systems of simultaneous equations to meet the initial conditions. To satisfy all these initial conditions, you end up with a series of simultaneous equations, one for $y(x_0) = y_0$, one for $y'(x_0) = y'_0$, and so on:

$$c_1 y_1 + c_2 y_2 + \ldots c_{n-1} y_{n-1} + c_n y_n = y_0$$

$$c_1 y'_1 + c_2 y'_2 + \ldots c_{n-1} y'_{n-1} + c_n y'_n = y'_0$$

$$c_1 y''_1 + c_2 y'_2 + \ldots c_{n-1} y''_{n-1} + c_n y''_n = y''_0$$

$$c_1 y^{(n-1)}{}_1 + c_2 y^{(n-1)}{}_2 + \ldots c_{n-1} y^{(n-1)}{}_{n-1} + c_n y^{(n-1)}{}_n = y^{(n-1)}{}_0$$

## A couple of cool theorems

In the following sections, I give you a couple of helpful theorems about higher order linear homogeneous differential equations. Put them to good use!

### The general solution of a higher order linear homogeneous equation

You may wonder whether the solution you've found to a homogeneous equation, $y = c_1y_1 + c_2y_2 + \ldots c_{n-1}y_{n-1} + c_ny_n$, is a general solution. In other words, you want to figure out whether it encompasses every solution of the differential equation. And that brings me to a theorem about the general solution of a higher order linear homogeneous differential equation.

**If you have *n* solutions, $y_1, y_2, \ldots y_n$, of a general linear homogeneous differential equation of order *n*:**

$$\frac{d^n y}{dx^n} + p_1(x)\frac{d^{n-1} y}{dx^{n-1}} + p_2(x)\frac{d^{n-2} y}{dx^{n-2}} + \ldots + p_{n-1}(x)\frac{dy}{dx} + p_n(x)\,y = 0$$

**then a linear combination of those solutions:**

$$y = c_1y_1 + c_2y_2 + \ldots c_{n-1}y_{n-1} + c_ny_n$$

**encompasses all solutions if $y_1, y_2, \ldots y_n$ are linearly independent.**

What does it mean to be *linearly independent?* Well, the functions $f_1, f_2, f_3 \ldots f_n$ are linearly *dependent* if there exists a set of constants $c_1, c_2, c_3 \ldots c_n$ (not all of which are zero) and an interval $I$ such that:

$$c_1f_1 + c_2f_2 + \ldots c_{n-1}f_{n-1} + c_nf_n = 0$$

for every $x$ in $I$. The $f_1, f_2, f_3 \ldots f_n$ are linearly *independent* if they aren't linearly *dependent.* It's as simple as that.

If you have *n* linearly independent solutions for a linear homogeneous differential equation of order *n*, you have a *fundamental set of solutions* for the differential equation. The general solution to the linear homogeneous differential equation is a linear combination of the functions in the fundamental set of solutions.

In other words, it's the same story as I discuss in Chapter 5 for second order linear homogeneous differential equations — only here it's generalized for higher orders.

### Solutions as related to the Wronskian

You can cast the theorem in the previous section in terms of the Wronskian, which is the determinant of this matrix for a higher order differential equation (see Chapter 5 for full details about the Wronskian):

$$W\left(y_1, y_2, \ldots y_n\right) = \begin{vmatrix} y_1 & y_2 & y_3 & \ldots & y_n \\ y'_1 & y'_2 & y'_3 & \ldots & y'_n \\ y''_2 & y''_2 & y''_3 & \ldots & y''_n \\ y^{(n-1)}{}_1 & y^{(n-1)}{}_2 & y^{(n-1)}{}_3 & \ldots & y^{(n-1)}{}_n \end{vmatrix}$$

Here's the theorem from the previous section written in terms of the Wronskian:

**If you have $n$ solutions, $y_1, y_2, \ldots y_n$, of a general linear homogeneous differential equation of order $n$:**

$$\frac{d^n y}{dx^n} + p_1(x)\frac{d^{n-1}}{dx^{n-1}} + p_2(x)\frac{d^{n-2}}{dx^{n-2}} + \ldots + p_{n-1}(x)\frac{dy}{dx} + p_n(x)\,y = 0$$

**and if their Wronskian, $W(y_1, y_2 \ldots y_n)(x) \neq 0$ in an interval $I$ for at least one point $x_0$ in that interval, then all solutions of the homogeneous equation are encompassed by linear combinations of those solutions.**

# Tackling Different Types of Higher Order Linear Homogeneous Equations

In the following sections, you can check out several different cases of higher order linear homogeneous equations: those with real and distinct roots, real and imaginary roots, complex roots, and duplicate roots.

## Real and distinct roots

The following sections walk you through one third order equation and one fourth order equation, both of which have real and distinct roots.

### A third order equation

Start by taking a look at the third order differential equation you solved at the beginning of this chapter:

$$y''' - 6y'' + 11y' - 6y = 0$$

Assume these initial conditions:

$$y(0) = 9$$
$$y'(0) = 20$$
$$y''(0) = 50$$

This differential equation has constant coefficients (see Chapter 5 for more information), so you can start by assuming a solution of the following form:

$$y = e^{rt}$$

Plugging this solution into your original equation gives you:

$$r^3e^{rt} - 6r^2e^{rt} + 11re^{rt} - 6e^{rt} = 0$$

Canceling out the $e^{rt}$ yields:

$$r^3 - 6r^2 + 11r - 6 = 0$$

The latter is the characteristic equation for the original homogeneous equation (check out Chapter 5 for more on characteristic equations). How do you solve this equation for the roots?

Here's when you start seeing one of the difficulties of higher order differential equations — they're like first and second order differential equations, only *more so.* That is, a second order differential equation gives you a second order characteristic equation, which you can solve with the quadratic equation. But what if you're faced with a differential equation of order 5? There is no "quintic" equation — you're on your own when it comes to finding the roots.

In the case of the characteristic equation in this example, you can (luckily) factor it into this:

$$(r - 1)(r - 2)(r - 3) = 0$$

So the roots are:

$$r_1 = 1$$
$$r_2 = 2$$
$$r_3 = 3$$

The roots are real and distinct, and the linearly independent solutions are:

$$y_1 = e^x$$
$$y_2 = e^{2x}$$
$$y_3 = e^{3x}$$

So the general solution to the original homogeneous equation is:

$$y = c_1e^x + c_2e^{2x} + c_3e^{3x}$$

Now turn to the initial conditions. In addition to the form for $y$, you also need $y'$ and $y''$ to meet the initial conditions:

$$y' = c_1e^x + 2c_2e^{2x} + 3c_3e^{3x}$$

and

$$y'' = c_1e^x + 4c_2e^{2x} + 9c_3e^{3x}$$

From the initial conditions, $y(0) = 9$, $y'(0) = 20$, and $y''(0) = 50$, here are your three simultaneous equations in $c_1$, $c_2$, and $c_3$:

$$y(0) = c_1 + c_2 + c_3 = 9$$
$$y'(0) = c_1 + 2c_2 + 3c_3 = 20$$
$$y''(0) = c_1 + 4c_2 + 9c_3 = 50$$

**WARNING!**

Once again, here's another place where you see the difference between first and second order differential equations and those of higher order. With first order differential equations that have initial conditions, solving for $c_1$ is trivial. With second order differential equations that have initial conditions, you end up with a 2 x 2 system of equations — and solving that is easy. However, starting with third order differential equations that have initial conditions, the $n$ x $n$ system of simultaneous equations can be a little more challenging.

**TIP**

Solving a 3 x 3 system of simultaneous equations for $c_1$, $c_2$, and $c_3$ involves some tedious algebra. You can calculate this out if you want to invest the time, or you can just use a computer. If you want to solve the system online, there are various services to do so. One such Web site is `math.cowpi.com/systemsolver`, which is a handy tool. You can simply select the type of system you're dealing with (such as 3 x 3, 4 x 4, and 5 x 5), plug in the right numbers, and voila! You have an answer.

The solutions of your equations, simply stated, are:

$$c_1 = 2$$
$$c_2 = 3$$
$$c_3 = 4$$

So the solution of the original homogeneous equation with the initial conditions applied is:

$$y = 2e^{x} + 3e^{2x} + 4e^{3x}$$

Cool. Good work!

### A fourth order equation

Now you're ready to try a differential equation of fourth order. Take a look at this beauty:

$$y^{(4)} + 10y''' + 35y'' + 50y + 24 = 0$$

Yep, that's a whopper. And it has initial conditions, too:

$$y(0) = 10$$

$$y'(0) = -20$$

$$y''(0) = 50$$

$$y'''(0) = -146$$

The differential equation has constant coefficients, so you can try a solution of the following form:

$$y = e^{rx}$$

Substituting this solution into your original homogeneous equation gives you:

$$r^4e^{rx} + 10r^3e^{rx} + 35r^2e^{rx} + 50re^{rx} + 24e^{rx} = 0$$

And dividing by $e^{rx}$ gives you this characteristic equation:

$$r^4 + 10r^3 + 35r^2 + 50r + 24 = 0$$

Okay, now you're in a pickle. What are the roots of this equation? Well, by just looking at it, you can figure out that it can be factored this way:

$$(r + 1)(r + 2)(r + 3)(r + 4) = 0$$

TIP

Just kidding — you can't *really* tell that just by looking at the equation; I only knew because I'm the one who made up the problem. So, if you have a characteristic equation like this one that's tough to factor, you can turn to online equation solvers. One of my favorites is `www.hostsrv.com/webmab/app1/MSP/quickmath/02/pageGenerate?site=quickmath&s1=equations&s2=solve&s3=basic` (yes, I know — that's a heck of an URL). A quick tip: When you enter variables raised to any power, be sure to add a caret (for instance, you enter $r^2$ as r^2). After you enter your equation, click the Solve button, and if your equation is factorable, you'll get the roots.

The roots of your characteristic equation are:

$$r_1 = -1$$

$$r_2 = -2$$

$$r_3 = -3$$

$$r_4 = -4$$

In other words, you have these three linearly independent solutions:

$$y_1 = e^{-x}$$

$$y_2 = e^{-2x}$$

$$y_3 = e^{-3x}$$

$$y_4 = e^{-4x}$$

So the general solution to the original homogeneous equation is:

$$y = c_1e^{-x} + c_2e^{-2x} + c_3e^{-3x} + c_4e^{-4x}$$

To meet the initial conditions, you need $y'$, $y''$, and $y'''$:

$$y' = -c_1e^{-x} -2c_2e^{-2x} - 3c_3e^{-3x} - 4c_4e^{-4x}$$

$$y'' = c_1e^{-x} + 4c_2e^{-2x} + 9c_3e^{-3x} + 16c_4e^{-4x}$$

$$y''' = -c_1e^{-x} - 8c_2e^{-2x} - 27c_3e^{-3x} - 64c_4e^{-4x}$$

Substituting the initial conditions gives you:

$$y(0) = c_1 + c_2 + c_3 + c_4 = 10$$

$$y'(0) = -c_1 - 2c_2 - 3c_3 - 4c_4 = -20$$

$$y''(0) = c_1 + 4c_2 + 9c_3 + 16c_4 = 50$$

$$y'''(0) = -c_1 - 8c_2 - 27c_3 - 64c_4 = -146$$

Well, this is another fine mess — you have a 4 x 4 system of simultaneous equations in $c_1$, $c_2$, $c_3$, and $c_4$. You can invest the time to solve it algebraically, or you can use a program to solve this system, such as `math.cowpi.com/systemsolver` (which I mention in the previous section).

You can see that:

$$c_1 = 4$$

$$c_2 = 3$$

$$c_3 = 2$$

$$c_4 = 1$$

which makes the general solution, including initial conditions, to the original homogeneous equation look like this:

$$y = 4e^{-x} + 3e^{-2x} + 2e^{-3x} + e^{-4x}$$

## Real and imaginary roots

Linear homogeneous equations with constant coefficients often have both real and imaginary roots. Check out this equation, for instance:

$$y^{(4)} - y = 0$$

Here are the initial conditions:

$$y(0) = 3$$

$$y'(0) = 1$$

$$y''(0) = -1$$

$$y'''(0) = -3$$

Because this is a linear homogeneous differential equation with constant coefficients, you decide to try a solution of the following form:

$$y = e^{rt}$$

Plugging the solution into the original equation gives you:

$$r^4e^{rt} - e^{rt} = 0$$

After canceling out $e^{rt}$, you get this equation:

$$r^4 - 1 = 0$$

As you may have noticed, this is a more manageable characteristic equation than the one in the previous section. You can easily factor this characteristic equation into:

$$(r^2 - 1)\,(r^2 + 1) = 0$$

So the roots of this characteristic equation are:

$$r_1 = 1$$

$$r_2 = -1$$

$$r_3 = i$$

$$r_4 = -i$$

It looks as though you have real and imaginary roots for $r_3$ and $r_4$. You can handle complex roots with the following relations:

$$e^{(\alpha + i\beta)x} = e^{\alpha x}(\cos \beta x + i \sin \beta x)$$

and:

$$e^{(\alpha - i\beta)x} = e^{\alpha x}(\cos \beta x - i \sin \beta x)$$

Here, $\alpha = 0$ for $r_3$ and $r_4$, and you get:

$$e^{i\beta x} = \cos \beta x + i \sin \beta x$$

and:

$$e^{-i\beta x} = \cos \beta x - i \sin \beta x$$

So $y_3$ and $y_4$ can be expressed as a linear combination of sines and cosines. Thus you have these solutions:

$$y_1 = e^x$$

$$y_2 = e^{-x}$$

$$y_3 = \cos x$$

$$y_4 = \sin x$$

The general solution to the original homogeneous equation is:

$$y = c_1 e^x + c_2 e^{-x} + c_3 \cos x + c_4 \sin x$$

TIP

What happened to the $i$ multiplying sin $\beta x$? As I discuss in Chapter 5, the $i$ can be absorbed into the constants $c_3$ and $c_4$, because $i$ is, after all, just a constant.

Now you have to handle the initial conditions. To do that, you also need $y'$, $y''$, and $y'''$:

$$y' = c_1e^x - c_2e^{-x} - c_3\sin x + c_4\cos x$$

$$y'' = c_1e^x + c_2e^{-x} - c_3\cos x - c_4\sin x$$

$$y''' = c_1e^x - c_2e^{-x} + c_3\sin x - c_4\cos x$$

Substituting the initial conditions into the equation gives you:

$$y(0) = c_1 + c_2 + c_3 = 3$$

$$y'(0) = c_1 - c_2 + c_4 = 1$$

$$y''(0) = c_1 + c_2 - c_3 = -1$$

$$y'''(0) = c_1 - c_2 - c_4 = -3$$

And there you have it again — a 4 x 4 system of simultaneous equations. You can solve this system by doing the algebra, or you can make it easy on yourself and check out an online simultaneous equation solver, such as `math.cowpi.com/systemsolver`. Go ahead; I'll wait for you to come up with the answers . . . .

Ready? So you have:

$$c_1 = 0$$

$$c_2 = 1$$

$$c_3 = 2$$

$$c_4 = 2$$

which means that the general solution with initial conditions is:

$$y = e^{-x} + 2\cos x + 2\sin x$$

It's lucky that $c_1 = 0$. Why? Otherwise the solution would grow exponentially. You can see a graph of the solution in Figure 7-1, where the exponential term dies away in time.

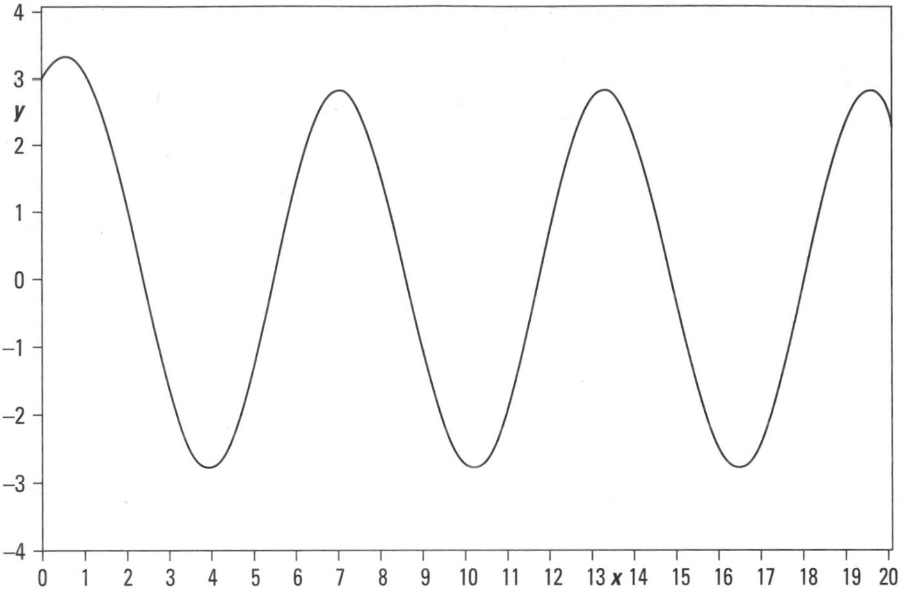


**Figure 7-1:** The solution to an equation with real and imaginary roots.

## Complex roots

The previous section handles the case where you have both real and imaginary roots. What about the case where you have complex roots? For example, take a look at this fine equation:

$$y^{(4)} + 4y = 0$$

Because this is a fourth order linear homogeneous differential equation with constant coefficients, you can try a solution of the following form:

$$y = e^{rx}$$

Plugging the solution into the original equation gives you:

$$r^4e^{rx} + 4e^{rx} = 0$$

Next, dividing by $e^{rx}$ gives you this characteristic equation:

$$r^4 + 4 = 0$$

So now you need the fourth root of –4, which isn't something you're likely to find on your calculator. Once again, these equations come to the rescue:

$$e^{(\alpha + i\beta)x} = e^{\alpha x}(\cos \beta x + i \sin \beta x)$$

and:

$$e^{(\alpha - i\beta)x} = e^{\alpha x}(\cos \beta x - i \sin \beta x)$$

You can think of –4 as $-4 + 0i$, so:

$$-4 = 4 \cos \pi + 4i \sin \pi = 4e^{i\pi}$$

Note that this relation is determined only up to multiples of $2\pi$, so this is actually:

$$-4 = 4 \cos (\pi + 2n\pi) + 4i \sin (\pi + 2n\pi) = 4e^{i(\pi + 2n\pi)}$$

where $n$ is an integer.

You can therefore find the fourth root of –4 this way:

$$(-4)^{1/4} = 4^{1/4}\, e^{i(\pi/4 + n\pi/2)}$$

Expanding the exponent gives you:

$$(-4)^{1/4} = 4^{1/4}\left(\cos (\pi/4 + n\pi/2) + i \sin (\pi/4 + n\pi/2)\right)$$

And using different values of $n$ gives you the fourth roots of –4:

$$r_1 = 1 + i$$

$$r_2 = -1 + i$$

$$r_3 = -1 - i$$

$$r_4 = 1 - i$$

Great! You've made progress. Now you know that the fundamental set of solutions is $e^{(1 + i)x}$, $e^{(-1 + i)x}$, $e^{(-1 - i)x}$, and $e^{(1 - i)x}$. By using these equations:

$$e^{(\alpha + i\beta)x} = e^{\alpha x}(\cos \beta x + i \sin \beta x)$$

and

$$e^{(\alpha - i\beta)x} = e^{\alpha x}(\cos \beta x - i \sin \beta x)$$

the fundamental set of solutions can be given this way in the general solution:

$$y = c_1 e^x \cos x + c_2 e^x \sin x + c_3 e^{-x} \cos x + c_4 e^{-x} \sin x$$

And that's the general solution of your original homogeneous equation. Cool.

## Duplicate roots

*Duplicate roots* are just what they sound like: one or more roots that are repeated as you figure out the general solution of a homogeneous equation. I describe several types of duplicate roots in the following sections.

### A fourth order equation with identical real roots

Take a look at this differential equation, which is a fourth order homogeneous differential equation with constant coefficients:

$$y^{(4)} + 4y''' + 6y'' + 4y' + y = 0$$

You can try a solution of the following form:

$$y = e^{rx}$$

Plugging the solution into the original homogeneous equation gives you this characteristic equation:

$$r^4 + 4r^3 + 6r^2 + 4r + 1 = 0$$

Holy mackerel, you might think. What are the roots of that thing? You can factor this equation using an online equation factoring program, such as `hostsrv.com/webmab/app1/MSP/quickmath/02/pageGenerate?site=quickmath&s1=equations&s2=solve&s3=basic`.

TIP

If you're determined to rely on your own brain to factor such an equation, here's a trick that sometimes helps: Convert the equation from base $r$ to base 10. That is, $r^2 + 2r + 1$ becomes $100 + 20 + 1 = 121$, which can be easily factored into 11 x 11. Converting 11 from base 10 back to base $r$ gives you $(r + 1)(r + 1)$, so the roots are –1 and –1. You can use this quick trick to astound your friends.

No matter which method you use, you'll sooner or later figure out that you can factor $r^4 + 4r^3 + 6r^2 + 4r + 1 = 0$ this way:

$$(r + 1)(r + 1)(r + 1)(r + 1)$$

So the roots of the characteristic equation are –1, –1, –1, –1 — all repeated real roots. Does that mean the solutions are:

$$y_1 = e^{-x}$$

$$y_2 = e^{-x}$$

$$y_3 = e^{-x}$$

$$y_4 = e^{-x}$$

REMEMBER

Hardly. You need linearly independent solutions to get a fundamental set of solutions from which to build your general solutions. And because these are all $e^{-x}$, they're obviously not linearly independent. (I talk about linear independence in more detail in the earlier section "The general solution of a higher order linear homogeneous equation.")

What can you do here? Well, clearly you can't have this as the general solution:

$$y = c_1e^{-x} + c_2e^{-x} + c_3e^{-x} + c_4e^{-x}$$

Why? Because that's really equivalent to:

$$y = ce^{-x}$$

where $c = c_1 + c_2 + c_3 + c_4$

TIP

So what can you do? Easy: You add powers of $x$ until you have as many linearly independent solutions as you need. You can convert $y = c_1e^{-x} + c_2e^{-x} + c_3e^{-x} + c_4e^{-x}$ into a true general solution by introducing factors of $x$, $x^2$, and $x^3$ like this:

$$y = c_1e^{-x} + c_2xe^{-x} + c_3x^2e^{-x} + c_4x^3e^{-x}$$

### A fifth order equation with identical real roots

Take a look at this whopper, which is a fifth order homogeneous differential equation:

$$y^{(5)} - y^{(4)} - 2y''' + 2y'' + y' - y = 0$$

Because it's homogeneous and has constant coefficients, you can try a solution in the form $y = e^{rx}$, giving you:

$$r^5e^{rx} - r^4e^{rx} - 2r^3e^{rx} + 2r^2e^{rx} + re^{rx} - e^{rx} = 0$$

After dividing by $e^{rx}$, you get this equation:

$$r^5 - r^4 - 2r^3 + 2r^2 + r - 1 = 0$$

Your response may sound something like this: "I'm supposed to factor that? Are you crazy?" Never fear; you can turn for help to an online equation solver, such as `hostsrv.com/webmab/app1/MSP/quickmath/02/pageGenerate?site=quickmath&s1=equations&s2=solve&s3=basic`.

The roots are –1, –1, 1, 1, 1, so $r^5 - r^4 - 2r^3 + 2r^2 + r - 1 = 0$ can be factored into this:

$$(r-1)(r-1)(r-1)(r+1)(r+1) = 0$$

or:

$$(r-1)^3(r+1)^2 = 0$$

As you can see, this equation looks a lot more manageable than $r^5 - r^4 - 2r^3 + 2r^2 + r - 1 = 0$. Because –1 is a root of the characteristic equation, this is a solution:

$$y_1 = e^{-x}$$

In fact, –1 is a double root, which also makes this a solution:

$$y_2 = xe^{-x}$$

Because you can also have 1 as a root, you have this solution:

$$y_3 = e^x$$

Finally, you may also notice that 1 is a triple root, so in that case you also have:

$$y_4 = xe^x$$

and

$$y_5 = x^2e^x$$

So the general solution to the original homogeneous equation is:

$$y = c_1e^{-x} + c_2xe^{-x} + c_3e^x + c_4xe^x + c_5x^2e^x$$

Notice that once again, the general solution is a linear combination of $n$ linearly independent solutions, where $n$ is the order of the differential equation. And note that also, you multiply solutions by $x$, $x^2$, and so on depending on their multiplicity as roots of the characteristic equation. Here, one root had multiplicity 2 and the other multiplicity 3.

### Identical imaginary roots

Are you curious about the case where the roots of a characteristic equation are duplicate and imaginary? For example, take a look at this differential equation with constant coefficients:

$$y^{(4)} + 8y'' + 16y = 0$$

You can try a solution of the following form:

$$y = e^{rx}$$

Substituting this solution into your original homogeneous equation gives you:

$$r^4e^{rx} + 8r^2e^{rx} + 16e^{rx} = 0$$

So your characteristic equation is:

$$r^4 + 8r^2 + 16 = 0$$

You can factor this into:

$$(r^2 + 4)(r^2 + 4) = 0$$

So the roots are $2i$, $2i$, $-2i$, and $-2i$. As you can see, you have duplicate imaginary roots here.

You can use these equations as I described earlier in this chapter:

$$e^{(\alpha + i\beta)x} = e^{\alpha x}(\cos \beta x + i \sin \beta x)$$

and:

$$e^{(\alpha - i\beta)x} = e^{\alpha x}(\cos \beta x - i \sin \beta x)$$

Here are the first two solutions, $y_1$ and $y_2$:

$$y_1 = \cos 2x$$

and

$$y_2 = \sin 2x$$

What about $y_3$ and $y_4$? You can use the same technique as in the previous section: Multiply by progressively higher powers of $x$ until you get all the linearly independent solutions you need. In this case, all you need is one power of $x$ to give you:

$$y_3 = x \cos 2x$$

and

$$y_4 = x \sin 2x$$

So the general solution to the homogeneous equation is:

$$y = c_1 \cos 2x + c_2 \sin 2x + c_3 x \cos 2x + c_4 x \sin 2x$$

### Identical complex roots

Here's another example; it's a fourth order equation with constant coefficients:

$$y^{(4)} - 8y''' + 32y'' - 64y' + 64y = 0$$

This equation is no problem. You just assume a solution of the form $y = e^{rx}$ and plug it into the original equation to get:

$$r^4e^{rx} - 8r^3e^{rx} + 32r^2e^{rx} - 64re^{rx} + 64e^{rx} = 0$$

So your characteristic equation is:

$$r^4 - 8r^3 + 32r^2 - 64r + 64 = 0$$

Solving this characteristic equation gives you these roots:

$$r_1 = 2 + 2i$$

$$r_2 = 2 + 2i$$

$$r_3 = 2 - 2i$$

$$r_4 = 2 - 2i$$

So $2 + 2i$ is a root of multiplicity 2, and so is $2 - 2i$. That leads to the following general solution:

$$y = b_1 e^{(2 + 2i)x} + b_2 x e^{(2 + 2i)x} + b_3 e^{(2 - 2i)x} + b_4 x e^{(2 - 2i)x}$$

where $b_1$, $b_2$, $b_3$, and $b_4$ are constants. You can rewrite this equation as the following:

$$y = e^{2x}(b_1 e^{2ix} + b_2 x e^{2ix}) + e^{2x}(b_3 e^{-2ix} + b_4 x e^{-2ix})$$

or:

$$y = e^{2x}(b_1 e^{2ix} + b_3 e^{-2ix}) + e^{2x}x(b_2 e^{2ix} + b_4 e^{-2ix})$$

Once again, you can turn to these relations:

$$e^{i\beta x} = \cos \beta x + i \sin \beta x$$

and:

$$e^{-i\beta x} = \cos \beta x - i \sin \beta x$$

Using these equations gives you:

$$y = e^{2x}\big(b_1(\cos 2x + i \sin 2x) + b_3 (\cos 2x - i \sin 2x)\big) + e^{2x}x(b_2(\cos 2x + i \sin 2x) + b_4(\cos 2x - i \sin 2x))$$

Combining terms and consolidating constants gives you this form for the general solution:

$$y = e^{2x}(c_1 \cos 2x + c_2 \sin 2x) + e^{2x}x(c_3 \cos 2x + c_4 \sin 2x)$$

And there you go, that's the general solution where the characteristic equation has the roots $2 + 2i$, $2 + 2i$, $2 - 2i$, and $2 - 2i$.

# Chapter 8

# Taking On Higher Order Linear Nonhomogeneous Differential Equations

## In This Chapter

- Breaking down higher order equations with the method of undetermined coefficients
- Using the variation of parameters to solve higher order equations

The door to your office opens. A group of rocket scientists enters, looking embarrassed. "We need you to solve a differential equation," they say.

"I thought you were math specialists," you retort.

"We thought so too. But this one has us stumped. It specifies the shape of the rocket. Without it, we can't take off." They slide a sheet of paper onto your desk. "Nobody has to know that we came to you for help, right?" They look over their shoulders nervously.

"Right," you say, taking a look at the sheet of paper, on which you see this differential equation:

$$y''' + 6y'' + 11y' + 6y = 336e^{5x}$$

The equation has the following initial conditions:

$$y(0) = 9$$

$$y'(0) = -7$$

$$y''(0) = 47$$

"It's a nonhomogeneous third order differential equation," the rocket scientists say. "You're stumped, right? We knew it!"

"Not so fast!" you say. "The solution is:"

$$y = 5e^{-x} + 2e^{-2x} + e^{-3x} + e^{5x}$$

The rocket scientists are stunned. "How did you do that?" they ask.

"Easy," you say. "I read Chapter 8 of *Differential Equations For Dummies.* I highly recommend it. Here's my bill for solving your equation."

The rocket scientists take the bill, and, reading it, raise their eyebrows. "That's astronomical," they say.

"Well, you *are* rocket scientists, aren't you?" you ask.

This chapter shows you how to handle nonhomogeneous linear differential equations of order $n$, where $n = 3, 4, 5$, and so on. They look like this:

$$\frac{d^n y}{dx^n} + p_1(x)\frac{d^{n-1}y}{dx^{n-1}} + p_2(x)\frac{d^{n-2}y}{dx^{n-2}} + \ldots + p_{n-1}(x)\frac{dy}{dx} + p_n(x)y = g(x)$$

In a higher order (or $n$th order) linear nonhomogeneous equation, the $p$ variables are various functions. And in this case, unlike in Chapter 7, where I discuss higher order homogeneous equations, $g(x) \neq 0$. When you're done with this chapter, you'll be able to handle nonhomogeneous equations with ease.

# Mastering the Method of Undetermined Coefficients for Higher Order Equations

I first discussed the *method of undetermined coefficients* in Chapter 6 for second order differential equations like this one:

$$y'' + p(x)y' + q(x)y = g(x)$$

In this chapter, however, I generalize that method to help you get a handle on differential equations of arbitrarily high order, not just two.

The method of undetermined coefficients is all about finding a particular solution to a nonhomogeneous equation, $y_p$. This method says that when you find a candidate solution, $y_p$, and plug it into the left-hand side of the equation, you end up with $g(x)$. Because $g(x)$ is only a function of $x$, it's often possible to guess the form of $y_p(x)$, up to arbitrary coefficients, and then solve for those coefficients by plugging $y_p(x)$ into the differential equation.

The form of $g(x)$ can often tell you what a particular solution looks like (just as it can when you're dealing with second order equations). In particular, if $g(x)$ is in the form of:

- **$e^{rx}$**, try a particular solution of the form $Ae^{rx}$, where $A$ is a constant. Because derivatives of $e^{rx}$ reproduce $e^{rx}$, you have a good chance of finding a particular solution this way.
- **A polynomial of order *n*,** try a polynomial of order $n$.
- **A combination of sines and cosines,** $\sin\alpha x + \cos\beta x$, try a combinations of sines and cosines with undetermined coefficients, $A\sin\alpha x + B\cos\beta x$. Then you can plug into the differential equation and solve for $A$ and $B$.

I explain these different forms in the following sections, but before I get to them, here's a summary of the method of undetermined coefficients for a higher order differential equation:

1. **Find the general solution, $y_h = c_1y_1 + c_2y_2 + \ldots + c_ny_n$ of the associated homogeneous differential equation.**
2. **If $g(x)$ is of the form $e^{rx}$, a polynomial, a combination of sines and cosines, or a product of any of these, assume that the particular solution is of the same form, using coefficients whose values have yet to be determined.**
3. **If $g(x)$ is the sum of terms, $g_1(x)$, $g_2(x)$, $g_3(x)$, and so on (as they are in a polynomial), break the problem into various subproblems like this:**

$$\frac{d^ny}{dx^n} + p_1(x)\frac{d^{n-1}y}{dx^{n-1}} + p_2(x)\frac{d^{n-2}y}{dx^{n-2}} + \ldots + p_{n-1}(x)\frac{dy}{dx} + p_n(x)y = g_1(x)$$

$$\frac{d^ny}{dx^n} + p_1(x)\frac{d^{n-1}y}{dx^{n-1}} + p_2(x)\frac{d^{n-2}y}{dx^{n-2}} + \ldots + p_{n-1}(x)\frac{dy}{dx} + p_n(x)y = g_2(x)$$

$$\frac{d^ny}{dx^n} + p_1(x)\frac{d^{n-1}y}{dx^{n-1}} + p_2(x)\frac{d^{n-2}y}{dx^{n-2}} + \ldots + p_{n-1}(x)\frac{dy}{dx} + p_n(x)y = g_3(x)$$

The particular solution of the nonhomogeneous equation is the sum of the solutions of those subproblems.

4. **Substitute $y_p$ into the differential equation, and solve for the undetermined coefficients.**
5. **Find the general solution of the nonhomogeneous differential equation, which is the sum of $y_h$ and $y_p$:**

$$y = y_h + y_p$$

6. **Use the initial conditions to solve for $c_1, c_2 \ldots c_n$.**

## When $g(x)$ is in the form $e^{rx}$

Take a look at the nonhomogeneous differential equation you solve so brilliantly for the rocket scientists at the beginning of this chapter:

$$y''' + 6y'' + 11y' + 6y = 336e^{5x}$$

Here are the equation's initial conditions:

$$y(0) = 9$$

$$y'(0) = -7$$

$$y''(0) = 47$$

REMEMBER

As with second order differential equations, the general solution to this nonhomgeneous differential equation is the sum of the solution to the corresponding homogeneous differential equation, in which $g(x)$ equals 0:

$$y''' + 6y'' + 11y' + 6y = 0$$

and a particular solution to the full nonhomogeneous differential equation. In other words:

$$y = y_h + y_p$$

where $y_h$ is the general solution to the homogeneous differential equation, and $y_p$ is a particular solution to the nonhomogeneous differential equation.

### The general solution to the homogeneous equation

The first order of business is to solve the homogeneous differential equation $y''' + 6y'' + 11y' + 6y = 0$. This is a third order linear homogeneous differential equation with constant coefficients (like in Chapter 7), so you decide to try a solution of the following form:

$$y = e^{rx}$$

Plugging this solution into the homogeneous equation gives you:

$$r^3e^{rx} + 6r^2e^{rx} + 11re^{rx} + 6e^{rx} = 0$$

And dividing by $e^{rx}$ (to make things a bit simpler) gives you the following characteristic equation (see Chapter 5 for more about this type of equation):

$$r^3 + 6r^2 + 11r + 6 = 0$$

If you're feeling extra sharp today and can do the algebra yourself, or if you have an online equation solver at your beck and call like the one at `www.hostsrv.com/webmab/app1/MSP/quickmath/02/pageGenerate?site=quickmath&s1=equations&s2=solve&s3=basic`, you can see that the roots of this equation are:

$$r_1 = -1$$

$$r_2 = -2$$

$$r_3 = -3$$

You can factor $r^3 + 6r^2 + 11r + 6 = 0$ this way:

$$(r + 1)(r + 2)(r + 3) = 0$$

So the general solution to the homogeneous differential equation $y''' + 6y'' + 11y' + 6y = 0$ is:

$$y_h = c_1e^{-x} + c_2e^{-2x} + c_3e^{-3x}$$

### *The particular solution to the nonhomogeneous equation*

After you find the general solution to the homogeneous equation, you have to find a particular solution, $y_p$, to the nonhomogeneous equation. In this case, $g(x)$ has the following form:

$$g(x) = 336e^{5x}$$

The method of undetermined coefficients says that you should try to match the form of $g(x)$ so that differentiating $y_p$ gives you the same form, up to the value of multiplicative constants. Because differentiating simple exponents gives you the same exponent back with possibly a different coefficient, the method of undetermined coefficients says that in this case you should try $y_p$ of this form:

$$y_p = Ae^{5x}$$

Substituting this solution into the equation gives you:

$$125Ae^{5x} + 150Ae^{5x} + 55Ae^{5x} + 6Ae^{5x} = 336e^{5x}$$

or:

$$125A + 150A + 55A + 6A = 336$$

Do the math and you get:

$$336A = 336$$

So:

$$A = 1$$

which means that the particular solution, $y_p$, of the nonhomogeneous equation is given by this:

$$y_p = e^{5x}$$

And because:

$$y = y_h + y_p$$

the general solution to the nonhomogeneous equation is:

$$y = c_1e^{-x} + c_2e^{-2x} + c_3e^{-3x} + e^{5x}$$

### Applying initial conditions

To handle the initial conditions of the original nonhomogeneous equation, you need to find $y$, $y'$ and $y''$:

$$y = c_1e^{-x} + c_2e^{-2x} + c_3e^{-3x} + e^{5x}$$
$$y' = -c_1e^{-x} - 2c_2e^{-2x} - 3c_3e^{-3x} + 5e^{5x}$$
$$y'' = c_1e^{-x} + 4c_2e^{-2x} + 9c_3e^{-3x} + 25e^{5x}$$

Plugging in $x = 0$ gives you the following:

$$y(0) = c_1 + c_2 + c_3 + 1 = 9$$
$$y'(0) = -c_1 - 2c_2 - 3c_3 + 5 = -7$$
$$y''(0) = c_1 + 4c_2 + 9c_3 + 25 = 47$$

Time to do a little simplifying:

$$y(0) = c_1 + c_2 + c_3 = 8$$

$$y'(0) = -c_1 - 2c_2 - 3c_3 = -12$$

$$y''(0) = c_1 + 4c_2 + 9c_3 = 22$$

This is a 3 x 3 system of simultaneous equations. You can work it out step by step, or you can use a handy equation system solver like the one at `math.cowpi.com/systemsolver`. Either way you work it, you find that $c_1$, $c_2$, and $c_3$ are:

$$c_1 = 5$$

$$c_2 = 2$$

$$c_3 = 1$$

So the general solution to your original nonhomogeneous equation with the given initial conditions is:

$$y = 5e^{-x} + 2e^{-2x} + e^{-3x} + e^{5x}$$

Just as you told the rocket scientists at the beginning of the chapter. Nice work!

## When g(x) is a polynomial of order n

Take a look at this differential equation stumper that I put together for you:

$$y''' - y' = x + 60e^{-4x} + 9 \sin x$$

It's the triple play: A third degree differential equation that's equal to a polynomial, an exponential, and a trig function — all at the same time. Seriously, though, solving this equation isn't as difficult as you may think; just check out the following sections.

### The general solution to the homogeneous equation

Start by looking for the general solution to the corresponding homogeneous differential equation:

$$y''' - y' = 0$$

This equation looks like a linear homogeneous differential equation with constant coefficients, so you can plug in the handy $y = e^{rx}$:

$$r^3e^{rx} - re^{rx} = 0$$

So the characteristic equation, after dropping $e^{rx}$, is:

$$r^3 - r = 0$$

One root is clearly $r = 0$, and dividing by $r$ gives you:

$$r^2 - 1 = 0$$

The roots of this equation are 1 and –1, and so the solution to the homogeneous differential equation is:

$$y_h = c_1 + c_2e^x - c_3e^{-x}$$

### *The particular solution to the nonhomogeneous equation*

Okay, so far so good. But now you have to find a particular solution, $y_p$. In this case, $g(x)$ is:

$$g(x) = x + 60e^{-4x} + 9 \sin x$$

Clearly, you have a polynomial on your hands. What to do? The way to handle this situation is to break $g(x)$ into three parts:

$$g_1(x) = x$$

$$g_2(x) = 60e^{-4x}$$

$$g_3(x) = 9 \sin x$$

Now you have three differential equations to solve:

$$y''' - y' = x$$

$$y''' - y' = 60e^{-4x}$$

$$y''' - y' = \sin x$$

And the corresponding particular solutions are $y_{p1}$, $y_{p2}$, and $y_{p3}$, respectively, which means that the general solution of the nonhomogeneous equation is:

$$y = y_h + y_{p1} + y_{p2} + y_{p3}$$

Alright, start by looking for $y_{p1}$, with $g_1(x) = x$. To do so you have to solve $y''' - y' = x$.

Because $g_1(x) = x$ here, you may think of using $y_{p1} = Ax$, but unfortunately, $x$ is already a solution of the homogeneous differential equation. So $y_{p1} = Ax$ is out. Instead, you can try:

$$y_{p1} = Ax^2$$

Plugging that solution into $y''' - y' = x$ gives you:

$$-2Ax = x$$

So:

$$A = -\tfrac{1}{2}$$

And therefore:

$$y_{p1} = Ax^2 = \frac{-x^2}{2}$$

Okay, now for $y_{p2}$, which is the solution to $y''' - y' = 60e^{-4x}$. Because $g_2(x) = e^{-4x}$ here, you can try a solution of the following form:

$$y_{p2} = Ae^{-4x}$$

Plugging $y_{p2}$ into $y''' - y' = 60e^{-4x}$ gives you:

$$-64Ae^{-4x} + 4Ae^{-4x} = 60e^{-4x}$$

or:

$$-60Ae^{-4x} = 60e^{-4x}$$

So:

$$A = -1$$

and

$$y_{p2} = -e^{-4x}$$

Great! You're making progress. The last thing you have to do is find $y_{p3}$, the third particular solution. That means solving $y''' - y' = \sin x$, for which you may be tempted to try a solution like this:

$$y_{p3} = A\cos x + B\sin x$$

However, take a look at the equation you're trying to solve here:

$$y''' - y' = \sin x$$

The first and third derivatives of $A\cos x$ yield terms in $\sin x$, which is what you're looking for here, but the first and third derivatives of $B\sin x$ yield terms in $\cos x$, which means that $B = 0$. So try a solution like $y_{p3} = A\cos x$. Plugging $y_{p3}$ into $y''' - y' = \sin x$ gives you:

$$A\sin x + A\sin x = \sin x$$

After dropping $\sin x$ you get:

$$A + A = 1$$

So:

$$A = \tfrac{1}{2}$$

which makes $y_{p3}$ equal to:

$$y_{p3} = \frac{\cos x}{2}$$

You now have all the particular solutions to the nonhomogeneous equation (and the general solution to the homogeneous equation from the previous section), so you can finally put everything together! Because:

$$y = y_h + y_{p1} + y_{p2} + y_{p3}$$

you get:

$$y = c_1 + c_2 e^{x} - c_3 e^{-x} - \frac{x^2}{2} - e^{-4x} + \frac{\cos x}{2}$$

## When g (x) is a combination of sines and cosines

Here's another higher order problem to work through; this time, the solution is a combination of sines and cosines, and the problem is of the fourth order:

$$y^{(4)} + 2y'' + y = 8\sin x - 16\cos x$$

Easy, right? All you have to do is find the general solution to the homogeneous equation, followed by the particular and general solutions to the nonhomogeneous equations. Read on for details.

### The general solution to the homogeneous equation

The homogeneous differential equation, in which $g(x)$ equals 0, is:

$$y^{(4)} + 2y'' + y = 0$$

This is a linear homogeneous differential equation with constant coefficients. This means you can try a solution of the form $y = e^{rx}$. Plugging this solution into the homogeneous equation gives you:

$$r^4e^{rx} + 2r^2e^{rx} + e^{rx} = 0$$

Or, for simplicity's sake:

$$r^4 + 2r^2 + 1 = 0$$

You can factor this into:

$$(r^2 + 1)\,(r^2 + 1) = 0$$

So the roots of the characteristic equation are $i$, $i$, $-i$, and $-i$. Two solutions to the homogeneous differential equation are:

$$y_1 = e^{ix}$$

and

$$y_2 = e^{-ix}$$

And because $i$ and $-i$ are repeated roots, you also have:

$$y_3 = x\,e^{ix}$$

and

$$y_4 = x\,e^{-ix}$$

So:

$$y_h = b_1e^{ix} + b_2e^{-ix} + b_3x\,e^{ix} + b_4\,x\,e^{-ix}$$

where $b_1 - b_4$ are constants. And because:

$$e^{i\beta x} = \cos \beta x + i \sin \beta x$$

and:

$$e^{-i\beta x} = \cos \beta x - i \sin \beta x$$

you can express the general solution to the homogeneous equation as:

$$y_h = c_1 \cos x + c_2 \sin x + c_3 x \cos x + c_4 x \sin x$$

### *The particular solution to the nonhomogeneous equation*

Now it's time to find a particular solution, $y_p$, of your original nonhomogeneous equation. In this case,

$$g(x) = 8 \sin x - 16 \cos x$$

You might consider trying a particular solution of the following form:

$$y_p = A \sin x + B \cos x$$

REMEMBER

But that particular solution isn't linearly independent from $y_h$, which has $\sin x$ and $\cos x$ terms in it. (Linear independence is important to find a complete set of solutions; see Chapter 5 for more information.) So you might consider a solution of the form $y_p = A\,x \sin x + B\,x \cos x$. But once again, this solution isn't linearly independent with respect to $y_h$, which has terms in $x \sin x + x \cos x$ already. So you're left with the following for $y_p$:

$$y_p = A\,x^2 \sin x + B\,x^2 \cos x$$

Substituting this solution into your nonhomogeneous equation, and collecting terms gives you:

$$-8\,A \sin x - 8\,B \cos x = 8 \sin x - 16 \cos x$$

So, comparing terms, you can see that:

$$A = -1$$

and:

$$B = 2$$

Now you know that the particular solution is:

$$y_p = -x^2 \sin x + 2\,x^2 \cos x$$

The general solution to the nonhomogeneous equation is:

$$y = c_1 \cos x + c_2 \sin x + c_3 x \cos x + c_4 x \sin x - x^2 \sin x + 2\,x^2 \cos x$$

Wow, a pretty lengthy solution. Impressive!

## A handy trick for finding particular solutions in sine and cosine form

The method of undetermined coefficients (which is discussed earlier in this chapter) is based on the fact that when $g(x)$ is of a certain form, you can often guess the form of the particular solution up to arbitrary coefficients. In fact, you can play a neat trick this way. If, for example, you have a sixth order homogeneous differential equation (yowza!) and know that $x^2e^{-x}\sin x$ is a solution, can you determine the other solutions? That is, given that:

$$y_1 = x^2 e^{-x} \sin x$$

can you find $y_2 - y_6$ so that all solutions are linearly independent? The short answer is this: Yes, you can. Because $x^2 e^{-x} \sin x$ is a solution, so are these two:

$$y_2 = e^{-x} \sin x$$

$$y_3 = x e^{-x} \sin x$$

Also, because $x^2 e^{-x} \sin x$ is a solution, so is $x^2 e^{-x} \cos x$, because complex roots of the characteristic equation come in conjugate pairs. This means that you can find the other three solutions:

$$y_4 = e^{-x} \cos x$$

$$y_5 = x e^{-x} \cos x$$

$$y_6 = x^2 e^{-x} \cos x$$

So the general solution to the homogeneous differential equation that has $x^2 e^{-x} \sin x$ as a solution is:

$$y = x^2 e^{-x} \sin x + e^{-x} \sin x + x e^{-x} \sin x + e^{-x} \cos x + x e^{-x} \cos x + x^2 e^{-x} \cos x$$

And you can determine all that starting with just one solution, $y_1 = x^2 e^{-x} \sin x$.

# Solving Higher Order Equations with Variation of Parameters

REMEMBER

The method of undetermined coefficients, which I discuss earlier in this chapter, is good only for certain forms of $g(x)$. For more general differential equations of a higher order, you can try the *method of variation of parameters*. This method was first introduced in Chapter 6 for differential equations of the second order, but here I generalize it for equations of order $n$.

## The basics of the method

To get a good grasp on this method, imagine that you have this general linear nonhomonegeous differential equation of order $n$:

$$\frac{d^n y}{dx^n} + p_1(x)\frac{d^{n-1} y}{dx^{n-1}} + p_2(x)\frac{d^{n-2} y}{dx^{n-2}} + \ldots + p_{n-1}(x)\frac{dy}{dx} + p_n(x)y = g(x)$$

The corresponding homogeneous differential equation is:

$$\frac{d^n y}{dx^n} + p_1(x)\frac{d^{n-1}y}{dx^{n-1}} + p_2(x)\frac{d^{n-2}y}{dx^{n-2}} + \ldots + p_{n-1}(x)\frac{dy}{dx} + p_n(x)y = 0$$

If the general solution to the homogeneous differential equation is:

$$y_h = c_1y_1 + c_2y_2 + c_3y_3 + \ldots + c_ny_n$$

then the variation of parameters says to look for a particular solution of the following form:

$$y_p = u_1(x)y_1 + u_2(x)y_2 + u_3(x)y_3 + \ldots + u_n(x)y_n$$

where $u_1(x)$, $u_2(x)$, and so on are functions.

REMEMBER

The method of variation of parameters for differential equations of order $n$ says that to find $y_p$, you can solve this system of simultaneous equations for $u'_1(x)$, $u'_2(x)$, and so on like this:

$$u'_1y_1 + u'_2y_2 + \ldots u'_ny_n = 0$$

$$u'_1y'_1 + u'_2y'_2 + \ldots u'_ny'_n = 0$$

$$u'_1y^{(n-2)}{}_1 + u'_2y^{(n-2)}{}_2 + \ldots u'_ny^{(n-2)}{}_n = 0$$

$$u'_1y^{(n-1)}{}_1 + u'_2y^{(n-1)}{}_2 + \ldots u'_ny^{(n-1)}{}_n = g(x)$$

Then you integrate $u'_1(x)$, $u'_2(x)$, and so on to find $u_1(x)$ and $u_2(x)$, which in turn gives you $y_p$.

## Working through an example

Here's an example that can help you get your feet wet with the method of variation of parameters. Try solving this differential equation using the method:

$$y^{(4)} = 6x$$

To do so, you first need the solution to the homogeneous differential equation:

$$y^{(4)} = 0$$

You can solve this equation by integration to get:

$$y_1 = 1$$

$$y_2 = x$$

$$y_3 = x^2$$

$$y_4 = x^3$$

The homogeneous equation's general solution is:

$$y_h = c_1 + c_2x + c_3x^2 + c_4x^3$$

Now you insert $u_1(x)$, $u_2(x)$, and so on for the constants to find the particular solution of the nonhomogeneous solution:

$$y_p = u_1(x) + u_2(x)x + u_3(x)x^2 + u_4(x)x^3$$

Here's where the method of variation of parameters kicks in, giving you these simultaneous equations in $u'_1(x)$, $u'_2(x)$, $u'_3(x)$, and $u'_4(x)$ by integration:

$$u'_1 + u'_2x + u'_3x^2 + u'_4x^3 = 0$$

$$u'_2 + u'_32x + u'_43x^2 = 0$$

$$u'_32 + u'_46x = 0$$

$$u'_46 = 6x$$

True, this is a set of four simultaneous equations in four unknowns, but it isn't so difficult to solve. Why? Because the equations get progressively simpler. Starting at the bottom, for example, you can see that after you cancel the 6 on each side, you get this:

$$u'_4(x) = x$$

You can then substitute that result into the previous equation:

$$u'_32 + u'_46x = 0$$

to get:

$$u'_32 + 6x^2 = 0$$

So:

$$u'_3(x) = -3x^2$$

You can find the others the same way. Here are $u'_1(x)$, $u'_2(x)$, $u'_3(x)$, and $u'_4(x)$:

$$u'_1 = -x^4$$

$$u'_2 = 3x^3$$

$$u'_3 = -3x^2$$

$$u'_4 = x$$

Integrating these gives you:

$$u_1 = \frac{-x^5}{5}$$

$$u_2 = \frac{3x^4}{4}$$

$$u_3 = -x^3$$

$$u_4 = \frac{x^2}{2}$$

Because:

$$y_p = u_1(x)y_1 + u_2(x)y_2 + u_3(x)y_3 + \ldots + u_n(x)y_n$$

the particular solution, $y_p$, equals:

$$y_p = \frac{-x^5}{5} + \frac{3x^5}{4} - x^5 + \frac{x^5}{2}$$

Or, to even out the denominators:

$$y_p = \frac{-4x^5}{20} + \frac{15x^5}{20} - \frac{20x^5}{20} + \frac{10x^5}{20}$$

So:

$$y_p = \frac{-4x^5}{20} + \frac{15x^5}{20} - \frac{20x^5}{20} + \frac{10x^5}{20}$$

which gives you the long-awaited result:

$$y_p = \frac{x^5}{20}$$

Now you know that the general solution of the nonhomogeneous equation is:

$$y_h = c_1 + c_2 x + c_3 x^2 + c_4 x^3 + \frac{x^5}{20}$$

And there you have it — you arrived at the general solution using the method of variation of parameters. (***Note:*** You can get the same result by simply integrating the differential equation four times.)

# Part III

# The Power Stuff: Advanced Techniques

The 5th Wave By Rich Tennant

"Great—differential equations brought us Newton's Law of Universal Gravitation, Maxwell's field equations, and now Stuart's Rate of Hair Loss."

## *In this part . . .*

This part is where I help you pull out the power tools. Here, you use series solutions, Laplace transforms, and systems of differential equations. In addition, you figure out how to use numerical methods to solve differential equations — this is the last-chance method, but it rarely fails.

# Chapter 9

# Getting Serious with Power Series and Ordinary Points

### *In This Chapter*

- Checking out the basics of power series
- Trying the ratio test
- Shifting the index value of a series
- Surveying the Taylor Series
- Putting your knowledge to use by solving second order equations

In Parts I and II of this book, I describe a variety of useful methods for solving first order, second order, and higher order differential equations. But sometimes, those methods, as cool as they are, just won't work. You have to solve some differential equations (such as those involving what differential equations experts call *ordinary points*) with a *power series* — that is, a summation of an infinite number of terms. I know this work sounds intimidating, but believe it or not, it's sometimes the easiest way to go. I show you what you need to know in this chapter.

## Perusing the Basics of Power Series

Power series are (often infinite) sums of terms. Here's an example of a common power series:

$$y = \sum_{n=0}^{\infty} a_n x^n$$

In this series, $a_n$ and $x^n$ are constants. The infinity sign on top of the sigma indicates that $n$ goes from 0 to infinity, and the sigma is the notation for a summation. This series is shorthand for the following infinite expansion, where the coefficients ($a_0$, $a_1$, $a_2$, and so on) are constants:

$$y = a_0 + a_1x + a_2x^2 + a_3x^3 + \ldots$$

The trouble with an infinite expansion, of course, is that it might *diverge.* That is, it might become infinite as you add more and more terms. Power series that become infinite aren't of much help to anyone, so in this chapter, you work only with series that stay finite — those that *converge* to a particular value. A power series is said to converge for a particular $x$ if this limit is finite:

$$\lim_{n \to \infty} \sum_{n=0}^{m} a_n x^n$$

If this limit is infinite, the series doesn't converge. In fact, your series might also converge *absolutely.* A series is said to converge absolutely if the summation of the absolute values of its terms converges (note the use of absolute value notation):

$$y = \sum_{n=0}^{\infty} \left| a_n x^n \right|$$

If a series converges absolutely, it also converges (of course).

# Determining Whether a Power Series Converges with the Ratio Test

So how do you know whether a series converges? That's an easy question: You use the *ratio test.* I discuss the basics of this test and provide a few numerical examples in the following sections.

## The fundamentals of the ratio test

The ratio test compares successive terms of a series to see whether the series will converge. If the ratio of the $(n + 1)$th term to the $n$th term is less than 1 for a fixed value of $x$, the series is said to converge for that $x$. The series diverges if the ratio is greater than 1. For example, suppose you had this series:

$$y = \sum_{n=0}^{\infty} a_n (x - x_0)^n$$

The ratio of the $(n + 1)$th term to the $n$th term is:

$$\frac{a_{n+1}(x-x_0)^{n+1}}{a_n(x-x_0)^n}$$

To find out whether the series converges or diverges, take a look at the limit of the ratio:

$$\lim_{n\to\infty}\frac{\left|a_{n+1}(x-x_0)^{n+1}\right|}{\left|a_n(x-x_0)^n\right|}$$

TIP

This limit can also be written this way:

$$\lim_{n\to\infty}\frac{\left|a_{n+1}(x-x_0)^{n+1}\right|}{\left|a_n(x-x_0)^n\right|}=\left|x-x_0\right|\lim\frac{\left|a_{n+1}\right|}{\left|a_n\right|}=\left|x-x_0\right|L$$

So the series is said to converge absolutely for a particular $x$ if $|x - x_0| < 1/L$. The series diverges if $|x - x_0| > 1/L$. And if $|x - x_0| = 1/L$, the ratio test is inconclusive and can't be used.

REMEMBER

In the large world of mathematics, there's a number that's either positive or zero, called the *radius of convergence*, $\rho$, such that the series converges absolutely if $|x - x_0| < \rho$ and diverges if $|x - x_0| > \rho$. The region in which $|x - x_0| < \rho$ in which the series converges is called the *interval of convergence.*

## Plugging in some numbers

The ratio test makes a lot more sense with numbers plugged into it. So in the following sections, I walk you through several examples. That way you can see for yourself just how useful this test is.

### Example 1

Here's an easy example to begin with. Where, if anywhere, does the following series converge absolutely?

$$\sum_{n=0}^{\infty}(-1)^n(x-3)^n$$

The first step is to look at the limit of the ratio of the $(n + 1)$th term to the $n$th term:

$$\lim_{n\to\infty}\frac{\left|(-1)^{n+1}(x-3)^{n+1}\right|}{\left|(-1)^n(x-3)^n\right|}$$

This ratio works out to:

$$\lim_{n\to\infty} \frac{\left|(-1)^{n+1}(x-3)^{n+1}\right|}{\left|(-1)^{n}(x-3)^{n}\right|} = |x-3|$$

As you can see, the ratio is $|x-3|$, and that ratio must be less than one. So the range in which the series converges absolutely is $|x-3| < 1$, or $2 < x < 4$. And the series diverges if $x < 2$ or $x > 4$.

### Example 2

Ready for another example? Take a look at this series:

$$\sum_{n=0}^{\infty} \frac{(x+1)^n}{4^n}$$

Determine the radius of convergence and the interval of convergence for this series. To do so, first apply the ratio test, which gives you this limit:

$$\lim_{n\to\infty} \left| \frac{4^n}{4^{n+1}} \frac{(x+1)^{n+1}}{(x+1)^n} \right| = \frac{|x+1|}{4}$$

As you can see, this series converges absolutely for $|x+1|/4 < 1$ or $|x+1| < 4$. So, in this case, the radius of convergence is 4, and the series converges for $-5 < x < 3$, which is the interval of convergence. That wasn't so difficult, was it?

### Example 3

Here's a final example showing how to use the ratio test. Take a look at this series:

$$\sum_{n=0}^{\infty} \frac{(2x+1)^n}{n^2}$$

As always, you use the ratio test, like so:

$$\lim_{n\to\infty} \frac{\left|(n+1)^2\right|}{\left|n^2\right|} \frac{\left|(2x+1)^{n+1}\right|}{\left|(2x+1)^n\right|} < 1$$

which equals:

$$|2x+1| \lim_{n\to\infty} \frac{\left|(n+1)^2\right|}{\left|n^2\right|} < 1$$

The limit evaluates to 1 as $n \to \infty$, so you get this:

$$|2x + 1| < 1$$

or:

$$|x + \tfrac{1}{2}| < \tfrac{1}{2}$$

The radius of convergence is ½, and the interval of convergence is $-1 < x < 0$.

# Shifting the Series Index

A method that can come in handy when working with differential equations is called *shifting the series index*. For example, say that you have this series, which starts at an index value of 3:

$$\sum_{n=3}^{\infty} a_n (x - x_0)^n$$

If you want this series to start at $n = 0$ instead of $n = 3$, you can simply shift the index by 3, like this:

$$\sum_{n=0}^{\infty} a_{n+3} (x - x_0)^{n+3}$$

Here's another example. Say that you have this series:

$$\sum_{n=2}^{\infty} (n+2)(x - x_0)^{n-2}$$

And say that you want to shift this so only powers of $n$ were involved, not $n - 2$. You can make this shift by replacing the dummy variable $n$ with $n + 2$, giving you:

$$\sum_{n=0}^{\infty} (n+4)(x - x_0)^n$$

# Taking a Look at the Taylor Series

You can express *continuous functions* (those functions that don't take discontinuous jumps) as a series: the *Taylor series.* The Taylor series says that a function can be expressed as an expansion around a point, $x_0$, like this:

$$f(x) = \sum_{n=0}^{\infty} \frac{f^{(n)}(x_0)}{n!}(x - x_0)$$

REMEMBER

In this series, $n!$ is $n$ factorial, or $n \cdot (n-1) \cdot (n-2) \ldots 3 \cdot 2 \cdot 1$. If a function has a Taylor series expansion at $x = x_0$ with a nonzero radius of convergence (see the earlier section "The fundamentals of the ratio test" for more about this radius), the function is said to be analytic at $x = x_0$.

A few types of Taylor series are especially important. Recognize this particular series?

$$\sum_{n=0}^{\infty} \frac{x^n}{n!} = e^x$$

It's your old friend $e^x$!

Here's sin $x$:

$$\sum_{n=0}^{\infty} \frac{(-1)^n x^{2n+1}}{(2n+1)!} = \sin x$$

And don't forget cos $x$:

$$\sum_{n=0}^{\infty} \frac{(-1)^n x^{2n}}{(2n)!} = \cos x$$

# Solving Second Order Differential Equations with Power Series

This section is all about tackling second order differential equations (which I introduce in Chapters 5 and 6) with power series. Say, for example, that you have a linear homogeneous second order differential equation like this:

$$P(x)\frac{d^2y}{dx^2} + Q(x)\frac{dy}{dx} + R(x)y = 0$$

TIP

Throughout the examples in this chapter, assume that $P(x)$, $Q(x)$, and $R(x)$ are all polynomials with no common factors. That's the easiest type of problem to solve with power series. However, this method is also applicable when $P(x)$, $Q(x)$, and $R(x)$ are general analytic functions, such as sin $x$ or cos $x$.

When working with power series, you divide the problems that fit the form of the previous equation into two types — those where you don't end up dividing by zero and those where you do. In this chapter, I focus on ordinary points. *Ordinary points* are points $x_0$ where $P(x_0)$ isn't equal to zero:

$$P(x_o) \neq 0$$

Because $P(x)$ is continuous, it follows that there's an interval around $x_0$ in which $P(x)$ isn't 0. Because $P(x)$ is a nonzero polynomial, you can divide by it to get the following (but remember that $P(x)$, $Q(x)$, and $R(x)$ have no common factors):

$$\frac{d^2y}{dx^2} + p(x)\frac{dy}{dx} + q(x)y = 0$$

where:

$$p(x) = \frac{Q(x)}{P(x)}$$

and:

$$q(x) = \frac{R(x)}{P(x)}$$

This equation is the type that you're going to look at in the following sections, and you're going to use ordinary points where nothing goofy happens (like functions suddenly going to infinity). In addition, you can impose initial conditions on this type of equation, such as:

$$y(0) = c_1$$

and

$$y'(0) = c_2$$

Points where functions go to infinity are called *singular points* (you can take a look at those in Chapter 10). At singular points, $P(x_0) = 0$, and at least one of $Q(x_0)$ and $R(x_0)$, isn't 0. So at least one of $p(x)$ or $q(x)$ becomes unbounded as $x \rightarrow x_0$. In this chapter, the functions are much better behaved.

So how do you solve second order differential equations using power series? The most basic way is to simply substitute a power series like this one:

$$\sum_{n=3}^{\infty} a_n(x - x_0)^n$$

into a differential equation. Then you can try to solve for the coefficients (which will be constants in this chapter). Because $p(x)$ and $q(x)$ are restricted to be polynomials in this chapter, working with power series will be even easier as you try to figure out the solution.

## When you already know the solution

Take a look at this second order differential equation:

$$\frac{d^2y}{dx^2} + y = 0$$

In your eagle-eyed solving mode, you no doubt recognize this equation to be a favorite among second order differential equations. You probably realize that the solution is:

$$y = c_1 y_1(x) + c_2 y_2(x)$$

where:

$$y_1(x) = \sin(x)$$

and

$$y_2(x) = \cos(x)$$

So you've solved the problem, but hold off crowing just yet, because you're going to tackle this problem using a power series. This problem is usually the first one you tackle, because you already know the solution, and you already know what sin($x$) and cos($x$) look like in terms of power series (see the earlier section "Taking a Look at the Taylor Series"). Here's sin($x$):

$$\sum_{n=0}^{\infty} \frac{(-1)^n x^{2n+1}}{(2n+1)!} = \sin x$$

And here's cos $x$:

$$\sum_{n=0}^{\infty} \frac{(-1)^n x^{2n}}{(2n)!} = \cos x$$

You're ready to start solving your original differential equation using a power series of the following form:

$$\sum_{n=0}^{\infty} a_n (x - x_0)^n$$

### Centering the power series on a selected point

REMEMBER

When solving a differential equation using a power series, you first have to select a point on which to center the power series ($x_0$ in the previous series). In the case of your original example equation, do what you'll usually do (unless there's a compelling reason otherwise): Choose $x_0$ to be 0. So your series will be an expansion of the solution around the ordinary point 0:

$$\sum_{n=0}^{\infty} a_n x^n$$

Take a deep breath and then substitute this series into the original differential equation you're trying to solve. To do so, you need to find the second derivative of the series. Keep reading to find out what to do.

### Finding the derivatives of a series

How do you find the second derivative of a series? As you may have guessed, you work term by term. So you start off with a solution $y$ of the following form:

$$y = \sum_{n=0}^{\infty} a_n x^n$$

To find $y''$, start by finding $y'$. Here's what the terms of the series look like:

$$y = a_0 + a_1x + a_2x^2 + a_3x^3 + \ldots$$

So after differentiating term by term you get

$$y' = a_1 + 2a_2x + 3a_3x^2 + \ldots$$

The general $n$th term here is:

$$na_nx^{n-1}$$

So $y'$ equals:

$$y' = \sum_{n=1}^{\infty} na_n x^{n-1}$$

TIP

Note that this series starts at $n = 1$, not $n = 0$ as the series for $y$ does, because you took the derivative of the series.

You can find $y''$ by differentiating $y' = a_1 + 2a_2x + 3a_3x^2 + \ldots$. You get this result for $y''$:

$$y'' = 2a_2 + 6a_3x + \ldots$$

The general term here is:

$$n(n-1)a_n x^{n-2}$$

So you can give $y''$ as the following:

$$y'' = \sum_{n=2}^{\infty} n(n-1)\,a_n x^{n-2}$$

### Substituting power series into the differential equation

The original differential equation looks like this:

$$\frac{d^2y}{dx^2} + y = 0$$

So you can substitute the power series for $y$ and $y''$ to get this result:

$$\sum_{n=2}^{\infty} n(n-1)\,a_n x^{n-2} + \sum_{n=0}^{\infty} a_n x^n = 0$$

which is your differential equation in series form.

### Ensuring the same index value

To compare the series in the differential equation, make sure they start at the same index value, $n = 0$. You can shift the first series here by replacing $n$ with $n + 2$ to get this result (I explain how to shift a series index in detail earlier in this chapter):

$$\sum_{n=0}^{\infty} (n+2)(n+1)\,a_{n+2} x^n + \sum_{n=0}^{\infty} a_n x^n = 0$$

When you do some simplifying, you get:

$$\sum_{n=0}^{\infty} \left[(n+2)(n+1)\,a_{n+2} x^n + a_n x^n\right] = 0$$

You can factor out $x^n$ as well:

$$\sum_{n=0}^{\infty} \left[(n+2)(n+1)\,a_{n+2} + a_n\right] x^n = 0$$

Believe it or not, you're making progress here. Seriously!

### Using the recurrence relation to find even coefficients

REMEMBER

Because the series shown in the previous section equals 0, and because it must work for all $x$, each term must equal 0. In other words, you get this:

$$(n+2)(n+1)a_{n+2} + a_n = 0$$

which is called a *recurrence relation.* When you have a differential equation's recurrence relation, you practically have the solution in your pocket. Alright, not really, but the recurrence relation goes a long way.

In this case, the recurrence relation says that, for $n = 0$:

$$(2)(1)a_2 + a_0 = 0$$

So:

$$a_2 = \frac{-a_0}{(2)(1)}$$

and for $n = 2$:

$$(4)(3)a_4 + a_2 = 0$$

So:

$$a_4 = \frac{-a_2}{(4)(3)}$$

Substituting $a_2$ from the first equation into the second equation gives you this result:

$$a_4 = \frac{a_0}{(4)(3)(2)(1)}$$

But $(4)(3)(2)(1) = 4!$, so you get:

$$a_4 = \frac{a_0}{4!}$$

Similarly, for $a_6$:

$$(6)(5)a_6 + a_4 = 0$$

or:

$$a_6 = \frac{-a_4}{(6)(5)}$$

Substituting for $a_4$ in terms of $a_0$ gives you:

$$a_6 = \frac{-a_0}{(6)(5)(4!)}$$

But $(6)(5)(4!) = 6!$, so you have:

$$a_6 = \frac{-a_0}{6!}$$

To summarize, you have the following:

$$a_2 = \frac{-a_0}{2!}$$

$$a_4 = \frac{a_0}{4!}$$

$$a_6 = \frac{-a_0}{6!}$$

So now you can relate the even coefficients in general. If $n = 2m$ (in which $m$ stands for a positive integer or zero):

$$a_n = a_{2m} = \frac{(-1)^m a_0}{(2m)!} \qquad m = 0, 1, 2, 3 \ldots$$

Because you set $a_0$ based on the initial conditions for a given problem, you can now find the even coefficients of the solution.

### Using the recurrence relation to find odd coefficients

Now you can move on to the odd coefficients. Turn back to the recurrence relation in the previous section for the solution:

$$(n + 2)(n + 1)a_{n+2} + a_n = 0$$

You can see that for $n = 1$ you get:

$$(3)(2)a_3 + a_1 = 0$$

So:

$$a_3 = \frac{-a_1}{(3)(2)}$$

Taking a clue from the even coefficients in the previous section, you can see that $(3)(2) = 3!$, so you get this:

$$a_3 = \frac{-a_1}{3!}$$

Now, for $n = 3$ in the recurrence relation, you get:

$$(5)(4)a_5 + a_3 = 0$$

or:

$$a_5 = \frac{-a_3}{(5)(4)}$$

Substituting $a_3$ in terms of $a_1$ into this equation gives you:

$$a_5 = \frac{a_1}{(5)(4)(3!)}$$

or:

$$a_5 = \frac{a_1}{5!}$$

Substituting $n = 5$ into the recurrence relation gives you:

$$(7)(6)a_7 + a_5 = 0$$

or:

$$a_7 = \frac{-a_5}{(7)(6)}$$

And substituting $a_5$ into this equation gives you:

$$a_7 = \frac{-a_1}{(7)(6)(5!)}$$

which means that:

$$a_7 = \frac{-a_1}{7!}$$

To sum up, the odd coefficients you have so far are:

$$a_3 = \frac{-a_1}{3!}$$

$$a_5 = \frac{a_1}{5!}$$

$$a_7 = \frac{-a_1}{7!}$$

So now you can relate the odd coefficients of the solution this way in general, if $n = 2m + 1$:

$$a_n = a_{2m+1} = \frac{(-1)^m a_1}{(2m+1)!} \qquad m = 0, 1, 2, 3 \ldots$$

### Putting together the solution

Using the two equations for even and odd coefficients from the previous sections, you get the following general solution to your differential equation in series terms:

$$y = a_0 \sum_{m=0}^{\infty} \frac{(-1)^m x^{2m}}{(2m)!}$$
$$+ a_1 \sum_{m=0}^{\infty} \frac{(-1)^m x^{2m+1}}{(2m+1)!}$$

That's the solution to the differential equation in series terms. In this case, the two series are recognizable as cos($x$) and sin($x$). Here's sin($x$):

$$\sum_{n=0}^{\infty} \frac{(-1)^n x^{2n+1}}{(2n+1)!} = \sin x$$

And here's cos $x$:

$$\sum_{n=0}^{\infty} \frac{(-1)^n x^{2n}}{(2n)!} = \cos x$$

So you can rewrite the solution as:

$$y = a_0 \cos(x) + a_1 \sin(x)$$

REMEMBER

The terms $a_0$ and $a_1$ are arbitrary constants (just like $c_0$ and $c_1$), and they're set by matching the initial conditions.

## When you don't know the solution beforehand

How do you use a power series to solve a differential equation for which you don't already know the solution? For instance, what if you have an equation like this:

$$\frac{d^2y}{dx^2} - x\frac{dy}{dx} + 2y = 0$$

Give it a try, using a power series like this:

$$y = \sum_{n=0}^{\infty} a_n x^n$$

The following sections will guide you through the process.

### Differentiating and substituting power series into the differential equation

Differentiating the power series from the previous section gives you the following equation:

$$y' = \sum_{n=1}^{\infty} n a_n x^{n-1}$$

And differentiating this gives you:

$$y''=\sum_{n=2}^{\infty} n(n-1)\,a_n x^{n-2}$$

Now for the fun part: substitution! Substituting these equations into the original differential equation gives you this result:

$$\sum_{n=2}^{\infty} n(n-1)\,a_n x^{n-2}$$

$$-x\sum_{n=1}^{\infty} na_n x^{n-1}$$

$$+2\sum_{n=0}^{\infty} a_n x^n = 0$$

### *Using the recurrence relation to find coefficients*

Now it's time to equate powers of $x$ on the left side of the equation to 0 on the right side. Combining the coefficients for powers of $x$ gives you this equation:

$$[2a_2 + 2a_0] + [6a_3 + a_1]\,x + 12a_4x^2 + [20a_5 - a_3]\,x^3 \ldots$$

$$+ [(n + 2)(n + 1)\,a_{n+2} - (n - 2)a_n\,]\,x^n = 0$$

Note that every power of $x$ must be 0 for the equation to work, so the last term in the previous equation gives you the recurrence relation for this solution:

$$(n + 2)(n + 1)\,a_{n+2} - (n - 2)a_n = 0$$

After some simplifying, you get:

$$a_{n+2}=\frac{(n-2)\,a_n}{(n+2)(n+1)}$$

That doesn't look so bad. This equation gives you these values for the coefficients when you plug in for $n$:

$$a_2 = -a_0$$

$$a_3=-\frac{1}{6}\,a_1$$

$$a_4 = 0$$

$$a_5=\frac{1}{20}\,a^3$$

Substituting $a_3$ in terms of $a_1$ into the previous equation gives you the following:

$$a_5=\frac{1}{20}\,a_3=-\frac{1}{120}\,a_1$$

Here's $a_6$:

$$a_6 = \frac{2}{30} a_4$$

And substituting $a_4$ into this equation gives you:

$$a_6 = \frac{2}{30} a_4 = \frac{2}{30}(0) = 0$$

Keep going! Here's $a_7$:

$$a_7 = \frac{3}{42} a_5$$

Or, to simplify:

$$a_7 = \frac{1}{14} a_5$$

Substituting $a_5$ in terms of $a_1$ into this equation gives you:

$$a_7 = \frac{1}{14} a_5 = \frac{1}{14} \frac{(-1)}{120} a_1$$

So:

$$a_7 = \frac{-1}{1{,}680} a_1$$

How about $a_8$? Here's what you get:

$$a_8 = \frac{4}{56} a_6$$

But you know that $a_6 = 0$, so:

$$a_8 = \frac{4}{56} a_6 = 0$$

In fact, because $a_4 = 0$, all subsequent even coefficients must be 0 by the recurrence relation:

$$a_{n+2} = \frac{(n-2)a_n}{(n+2)(n+1)}$$

### *Putting together the solution*

Because all the even coefficients are 0 beyond $a_2$, life is a little easier. You just need to plug in the odd coefficients to get the entire solution to the differential equation. Here's what you have for the solution:

$$y = a_0 + a_1 x - a_0 x^2 - \frac{1}{6} a_1 x^3 - \frac{1}{120} a_1 x^5 - \frac{1}{1{,}680} a_1 x^7 + \ldots$$

Grouping terms together by their coefficients gives you:

$$y = a_0(1 - x) +$$
$$a_1\left(x - \frac{1}{6}a_1x^3 - \frac{1}{120}a_1x^5 - \frac{1}{1{,}680}a_1x^7 + \ldots\right)$$

If you define the following:

$$y_1 = (1 - x^2)$$

and:

$$y_2 = \left(x - \frac{1}{6}a_1x^3 - \frac{1}{120}a_1x^5 - \frac{1}{1{,}680}a_1x^7 + \ldots\right)$$

then you can see that the general solution is:

$$y = a_0y_1 + a_1y_2$$

## A famous problem: Airy's equation

Before you wrap up the chapter, take a look at one more differential equation. Get ready, though. It's a famous one! In fact, it even has its own name — Airy's equation. Here's what it looks like:

$$y'' - xy - 0$$

To solve this equation, you can assume, as you do earlier in this chapter, that the solution is of the following form:

$$y = \sum_{n=0}^{\infty} a_n x^n$$

I explain the rest of the process in the following sections.

### Differentiating and substituting power series into the differential equation

As always, be sure to start off by differentiating the equation. Doing so gives you:

$$y' = \sum_{n=1}^{\infty} na_n x^{n-1}$$

After differentiating this equation one more time, you get:

$$y'' = \sum_{n=2}^{\infty} n(n-1)a_n x^{n-2}$$

## Who was Airy?

Sir George Biddell Airy (1801–1892) was an English mathematician and astronomer. In fact, he was the Astronomer Royal from 1835 to 1881. He was a man of many achievements. For instance, he found planetary orbits, measured the Earth's density, and came up with a solution of two-dimensional problems in mechanics. He's also the one responsible for establishing Greenwich in Britain at the location of the prime meridian.

Plugging the derivatives into the original differential equation gives you this result:

$$\sum_{n=2}^{\infty} n(n-1)\,a_n x^{n-2} - x\sum_{n=0}^{\infty} a_n x^n = 0$$

With some simplifying you get:

$$\sum_{n=2}^{\infty} n(n-1)\,a_n x^{n-2} - \sum_{n=0}^{\infty} a_n x^{n+1} = 0$$

You can also write this equation as:

$$\sum_{n=2}^{\infty} n(n-1)\,a_n x^{n-2} - \sum_{n=0}^{\infty} a_n x^{n+1}$$

### *Ensuring the same index value*

Now you need to compare the coefficients of equal powers of $x$ on the two sides of this equation to find the recurrence relationship. To compare those coefficients, it helps to make sure that the powers of $x$ are the same in each series.

Compare the coefficients in terms of $x^n$. Shifting the series on the right (as I show you earlier in this chapter) gives you this result:

$$\sum_{n=2}^{\infty} n(n-1)\,a_n x^{n-2} - \sum_{n=1}^{\infty} a_{n-1} x^n$$

You can also shift the series on the left by substituting $n + 2$ for $n$, which gives you:

$$\sum_{n=0}^{\infty} (n+2)(n+1)\,a_{n+2} x^n = \sum_{n=1}^{\infty} a_{n-1} x^n$$

### *Using the recurrence relation to find coefficients*

TIP

Now you're ready to compare coefficients of powers of $x$. Note from the previous section that the series on the left starts at $n = 0$ and the series on the right starts at $n = 1$; to compare the terms, it's easier if the series starts at the same value, say $n = 1$. You can break the first term out of the series on the left to make the initial values of the indexes of the two series match, like this:

$$2a_2 + \sum_{n=1}^{\infty}(n+2)(n+1)a_{n+2}x^n = \sum_{n=1}^{\infty}a_{n-1}x^n$$

Note that the left side of this equation has a constant term, $2a_2$, but the right side doesn't. So right off the bat you know that $a_2 = 0$.

Because the two sides of this equation must be equal in their terms, you get the following for the recurrence relation:

$$(n + 2)(n + 1)a_{n+2} = a_{n-1} \quad n = 1, 2, 3, 4 \ldots$$

REMEMBER

The coefficients are determined in steps of three because if you know $a_{n-1}$, you can get $a_{n+2}$. That is, when you have $a_0$, you know you can get $a_3$, $a_6$, $a_9$, and so on. And when you have $a_1$, you can get $a_4$, $a_7$, $a_{10}$, and so on. Because you already know that $a_2 = 0$, you know that $a_5 = 0$, $a_8 = 0$, $a_{11} = 0$, and so on.

Start with the sequence $a_0$, $a_3$, $a_6$, $a_9$, and so on. And remember that $a_0$ and $a_1$ are set by the initial conditions. This means that you should start with $a_3$. Here's $a_3$ in terms of $a_0$:

$$a_3 = \frac{a_0}{(2)(3)}$$

Here's $a_6$:

$$a_6 = \frac{a_3}{(5)(6)}$$

Substituting $a_3$ in terms of $a_0$ into this equation gives you:

$$a_6 = \frac{a_0}{(5)(6)(2)(3)}$$

Here's $a_9$:

$$a_9 = \frac{a_6}{(8)(9)}$$

Substituting $a_6$ into this equation gives you:

$$a_9 = \frac{a_0}{(8)(9)(5)(6)(2)(3)}$$

So:

$$a_{3n} = \frac{a_0}{(2)(3)(5)(6)\ldots(3n-4)(3n-3)(3n-1)(3n)} \qquad n = 1, 2, 3\ldots$$

Now look at the sequence $a_1$, $a_4$, $a_7$, $a_{10}$, and so on. Because $a_1$ is set by the initial conditions, you should start with $a_4$:

$$a_4 = \frac{a_1}{(3)(4)}$$

Here's $a_7$:

$$a_7 = \frac{a_4}{(6)(7)}$$

Substituting $a_4$ in terms of $a_1$ gives you:

$$a_7 = \frac{a_1}{(6)(7)(3)(4)}$$

And here's $a_{10}$:

$$a_{10} = \frac{a_7}{(9)(10)}$$

Substituting $a_7$ into this equation gives you:

$$a_{10} = \frac{a_1}{(9)(10)(6)(7)(3)(4)}$$

So you can say that in general:

$$a_{3n+1} = \frac{a_1}{(3)(4)(6)(7)\ldots(3n-3)(3n-2)(3n)(3n+1)} \qquad n = 1, 2, 3\ldots$$

### *Putting together the solution*

After you get through with all the previous steps, you'll know that you can write the general solution to Airy's equation like this:

$$y = a_0\left[1 + \frac{x^3}{6} + \frac{x^6}{180} + \ldots \frac{x^{3n}}{(2)(3)(5)(6)\ldots(3n-4)(3n-3)(3n-1)(3n)}\right] +$$
$$a_1\left[x + \frac{x^4}{12} + \frac{x^7}{504} + \ldots \frac{x^{3n+1}}{(3)(4)(6)(7)\ldots(3n-3)(3n-2)(3n)(3n+1)}\right]$$

You can write this in the following shorter form:

$$y = a_o\left[1 + \sum_{n=0}^{\infty} \frac{x^{3n}}{(2)(3)(5)(6)\ldots(3n-4)(3n-3)(3n-1)(3n)}\right] +$$
$$a_1\left[\sum_{n=0}^{\infty} \frac{x^{3n+1}}{(3)(4)(6)(7)\ldots(3n-3)(3n-2)(3n)(3n+1)}\right]$$

So if you write:

$$y_1 = 1 + \sum_{n=0}^{\infty} \frac{x^{3n}}{(2)(3)(5)(6)\ldots(3n-4)(3n-3)(3n-1)(3n)}$$

and:

$$y_2 = \sum_{n=0}^{\infty} \frac{x^{3n+1}}{(3)(4)(6)(7)\ldots(3n-3)(3n-2)(3n)(3n+1)}$$

you can see that the general solution to Airy's equation is:

$$y = a_0 y_1 + a_1 y_2$$

# Chapter 10

# Powering through Singular Points

### In This Chapter

- Perusing singular points
- Examining Euler equations
- Surveying series solutions near regular singular points

In Chapter 9, I briefly introduce you to the concept of singular points, but this chapter is where the real action is. I cover a number of topics, including working with Euler equations, handling power series solutions near singular points, and dealing with a mix of the two — series solutions to Euler equations near singular points.

## Pointing Out the Basics of Singular Points

In this chapter, you work with second order homogeneous differential equations of the following form:

$$P(x)\frac{d^2y}{dx^2}+Q(x)\frac{dy}{dx}+R(x)y=0$$

where $P(x)$, $Q(x)$, and $R(x)$ have no common factors.

You can also divide each term in the equation by $P(x)$ to get:

$$\frac{d^2y}{dx^2}+p(x)\frac{dy}{dx}+q(x)y=0$$

where $p(x) = Q(x)/P(x)$ and $q(x) = R(x)/P(x)$.

So far, so good, right? Not so fast! Now I'm going to throw singular points into the mix. Points where functions go to infinity are called *singular points.* At singular points, $P(x_0) = 0$, and at least one of $Q(x_0)$ and $R(x_0)$ isn't zero. So at least one of $p(x)$ or $q(x)$ becomes *unbounded* (goes to infinity) as $x \rightarrow x_0$.

In the following sections, I introduce you to the fundamentals of singular points, including how to find them, how they behave, and the difference between regular and irregular points.

## Finding singular points

Time to get your feet wet! What are the singular points of the following differential equation?

$$(4-x^2)\frac{d^2y}{dx^2}+x^3\frac{dy}{dx}+(1+x)y=0$$

The singular points are where $P(x) = 0$, so you have:

$$(4-x^2) = 0$$

Simply use your excellent algebra skills to solve this equation, and you find that the singular points are $x = \pm 2$.

As another example, determine the singular points of this differential equation:

$$x^2\frac{d^2y}{dx^2}+(8-x^3)\frac{dy}{dx}+(1-9x)y=0$$

Here, $P(x)$ is simply $x^2$. So the singular point is $x = 0$.

## The behavior of singular points

When studied closely, singular points look like they have the potential to become unruly. After all, singular points are where a solution may go to infinity or change rapidly in magnitude. If you're asking whether you can simply ignore them, the answer is no.

Why? Because solutions to differential equations with singular points vary so much near those singular points that you need to take special care. In fact, it's often the case that the solution to a differential equation is the most interesting around its singular points. That's where the most interesting physics goes on. For example, an electrical circuit may reach resonance there. And reaching resonance is often the whole point of amplifying circuits, so ignoring the behavior at singular points just wouldn't do.

For example, take a look at this second order homogeneous equation:

$$x^2\frac{d^2y}{dx^2} - 2y = 0$$

By using your unbeatable differential equation solving skills, you can tell that the two independent solutions to this differential equation are:

$$y_1 = x^2$$

and

$$y_2 = x^{-1}$$

The $y_1$ solution is fine — its behavior is well-defined around $x = 0$, for example. In fact, the $y_1$ solution is still analytic (meaning that it has a working Taylor expansion; see Chapter 9 for more about Taylor series) as $x \to 0$, even though the differential equation looks like this if you divide by $P(x)$:

$$\frac{d^2y}{dx^2} - \frac{2y}{x^2} = 0$$

This equation looks unbounded as $x \to 0$, but everything is okay if you substitute $y_1 = x^2$ here.

The situation is different with $y_2 = x^{-1}$ (also written as $1/x$), however. This solution isn't analytic at $x = 0$ (in other words, it has no Taylor expansion at $x = 0$). So you can't use the methods of Chapter 9 to solve this differential equation.

REMEMBER

As you find out in the next section (and throughout this chapter), different kinds of singular points exist — and all of them have varying levels of manageability. It is often the case that you can't use the Taylor expansion used in Chapter 9 to solve differential equations with singular points, because those expansions become unbounded at the singular points. Because of this, you have to use more general expansions from time to time.

## Regular versus irregular singular points

REMEMBER

An important concept when it comes to singular points is *severity*. A singular point's severity is an indication of how strong the singular point is — how strongly it tends toward infinity. The severity of a singular point has a lot to do with how you handle the solution; does the solution, for instance, include terms like $1/x$, or does it include terms like $1/x^8$? (It turns out that you can handle $1/x$ in most cases, but $1/x^8$? Good luck with that one.) The idea is to extend the techniques of Chapter 9, which allow you to use series expansions near ordinary points, to help you use series expansions near singular points — if they're well-behaved enough.

A well-behaved singular point is called a *regular singular point.* Regular singular points are well-behaved enough that you can handle them using the techniques in this chapter.

What's the definition of a regular singular point? They're defined in terms of the ratio $Q(x)/P(x)$ and $R(x)/P(x)$, where $P(x)$, $Q(x)$, $R(x)$ are the polynomial coefficients in the differential equation that you're trying to solve:

$$P(x)\frac{d^2y}{dx^2} + Q(x)\frac{dy}{dx} + R(x)y = 0$$

In order for $x_0$ to be a regular singular point of this equation, these two relations have to be true:

$$\lim_{x \to x_0} (x - x_0)\frac{Q(x)}{P(x)} \quad \text{remains finite}$$

and

$$\lim_{x \to x_0} (x - x_0)^2\frac{R(x)}{P(x)} \quad \text{remains finite}$$

Another way of thinking about this is to say that $x_0$ is a regular singular point if $(x - x_0)Q(x)/P(x)$ and $(x - x_0)^2Q(x)/P(x)$ have Taylor expansions (that is, they're analytic) around $x_0$.

As you may have guessed, if singular points aren't regular, they're irregular. And irregular singular points are much more — shall I say interesting? — to handle.

Take a look at some example differential equations to see if their singular points are regular or irregular.

### Example 1

Start with this differential equation (which you just saw in the previous section):

$$\frac{d^2y}{dx^2} - \frac{2y}{x^2} = 0$$

The solutions for this equation are:

$$y_1 = x^2$$

and

$$y_2 = x^{-1}$$

There's a singular point at $x = 0$ because $y_2$ becomes infinite. Now take some time to evaluate the relations:

$$\lim_{x \to x_0} (x - x_0) \frac{Q(x)}{P(x)} \text{ remains finite}$$

and

$$\lim_{x \to x_0} (x - x_0)^2 \frac{R(x)}{P(x)} \text{ remains finite}$$

You test the first relation by plugging in some numbers, like so:

$$\lim_{x \to 0} (x) \frac{0}{1} = 0$$

And because 0 is finite, you're okay so far. Now for the second relation, $R(x) = -2/x^2$ and $P(x) = 1$:

$$\lim_{x \to 0} (x)^2 \frac{-2}{x^2} = -2$$

Because –2 is finite, $x = 0$ is a regular singular point of your original differential equation. Cool, huh?

### Example 2

Now try this one:

$$(x-4)^2 x \frac{d^2y}{dx^2} + 4x \frac{dy}{dx} + (x-4)y = 0$$

What are the singular points in this equation? Are they regular? Here, $P(x) = (x - 4)^2 x$, and that's 0 at $x = 0$ and $x = 4$. So $x = 0$ and $x = 4$ are your singular points. Start with the $x = 0$ singular point first; here's what the first relation has to say about it:

$$\lim_{x \to 0} (x) \frac{4x}{(x-4)^2 x} = \frac{1}{4}$$

Because ¼ is finite, you're good to go. Now for the second relation:

$$\lim_{x \to 0} (x)^2 \frac{(x-4)}{(x-4)^2 x} = 0$$

Zero is finite, so the $x = 0$ singular point is a regular singular point.

So far, so good. Now, however, you have to try the other singular point, $x = 4$. The first relation gives you:

$$\lim_{x \to 4} (x-4) \frac{4x}{(x-4)^2 x} \rightarrow \frac{4}{(x-4)} \text{ is NOT FINITE}$$

Whoops — this relation is unbounded as $x \to 4$, so $x = 4$ isn't a regular singular point. In other words, it's an irregular singular point.

### Example 3

Try your hand at this equation, which is a famous one, the Legendre equation (see the nearby sidebar "Discovering the legacy of Legendre" for more on the man who unearthed this famous equation):

$$(1 - x^2)\frac{d^2y}{dx^2} - 2x\frac{dy}{dx} + \alpha(\alpha + 1)y = 0$$

where $\alpha$ is a constant. Because $P(x) = (1 - x^2)$, the singular points are $x = \pm 1$. Look at the $x = 1$ point first. From the first relation and because $(1 + x)(1 - x) = (1 - x^2)$:

$$\lim_{x \to 1}(x - 1)\frac{2x}{(1 - x^2)} = \frac{-2x}{(1 + x)} = -1$$

As you know, –1 is, of course, finite.

Now for the second relation:

$$\lim_{x \to 1}(x - 1)^2\frac{\alpha(\alpha + 1)}{(1 - x^2)}$$

This equation can be broken down to:

$$\lim_{x \to 1}(x - 1)^2\frac{\alpha(\alpha + 1)}{(1 - x)(1 + x)}$$

which is:

$$\lim_{x \to 1} -(x - 1)\frac{\alpha(\alpha + 1)}{(1 + x)} = 0$$

Because 0 is finite, $x = 1$ is a regular singular point.

Now try the other singular point, $x = -1$. Here's what the first relation gives you:

$$\lim_{x \to -1}(x + 1)\frac{2x}{(1 - x^2)} = \frac{2x}{(1 - x)} = -1$$

As you know, –1 is finite. Now check out the second relation, which gives you:

$$\lim_{x \to -1}(x + 1)^2\frac{\alpha(\alpha + 1)}{(1 - x^2)}$$

## Discovering the legacy of Legendre

Adrien-Marie Legendre (1752–1833) was the French mathematician who discovered the ever-important Legendre equation. His work was done in the fields of statistics, abstract algebra, mathematical analysis, and number theory. He was born to a rich family, and studied physics in Paris. Then he took up teaching at a military academy. He took on this job because he enjoyed it, not because he had to. He started publishing works on physics — specifically on the motion of cannonballs — but then moved on to math. In 1782, he became a member of the French Academy of Sciences. He even got a crater on the moon named after him. How many mathematicians can say that?

TIP

You can also write this as:

$$\lim_{x \to -1} (x+1)^2 \frac{\alpha(\alpha+1)}{(1-x)(1+x)}$$

or:

$$\lim_{x \to -1} (x+1) \frac{\alpha(\alpha+1)}{(1-x)}$$

The limit is:

$$\lim_{x \to -1} (x+1) \frac{\alpha(\alpha+1)}{(1-x)} = 0$$

Because 0 is finite, $x = -1$ is also a regular singular point.

# Exploring Exciting Euler Equations

A good way to understand how to handle regular singular points is to see how one of the most famous differential equations — which has a regular singular point — is handled. The equation I'm talking about is Euler's equation:

$$x^2 \frac{d^2 y}{dx^2} + \alpha x \frac{dy}{dx} + \beta y = 0$$

Here, α and β are real constants. If you assume that the solution to this equation has the following form:

$$y = x^r$$

(where $r$ is a constant), substituting the solution into the equation gives you:

$$[r(r-1) + \alpha r + \beta]\, x^r = 0$$

Dividing both sides by $x^r$ gives you:

$$r(r-1) + \alpha r + \beta = 0$$

or:

$$r^2 - r + \alpha r + \beta = 0$$

So:

$$r^2 + (\alpha - 1)\, r + \beta = 0$$

The roots, $r_1$ and $r_2$, of this equation are:

$$r_1, r_2 = \frac{-(\alpha - 1) \pm \sqrt{(\alpha - 1)^2 - 4\beta}}{2}$$

Because you're considering the general Euler's equation, you don't know what α and β are, so you have to consider three cases for these roots:

- $r_1$ and $r_2$ are real and distinct
- $r_1$ and $r_2$ are real and equal
- $r_1$ and $r_2$ are complex conjugates

In the following section, you consider each of these cases separately.

TIP

Euler was a pretty busy guy; he also came up with Euler's method, which is a simple numerical method of solving differential equations. Flip to Chapter 4 for details.

## Real and distinct roots

REMEMBER

If the roots of $r^2 + (\alpha - 1)\, r + \beta = 0$ are real and distinct, $r_1 \neq r_2$. The general solution to Euler's equation is:

$$y = c_1 x^{r_1} + c_2 x^{r_2}$$

To see how to come up with this solution, try solving this differential equation:

$$4x^2 \frac{d^2y}{dx^2} + 6x \frac{dy}{dx} - 2y = 0$$

This equation has a regular singular point at $x = 0$, and you can assume that the solution is of the following form:

$$y = x^r$$

Substituting this solution into the original equation gives you:

$$[4r(r-1) + 6r - 2]\, x^r = 0$$

or:

$$[4r(r-1) + 6r - 2] = 0$$

After some simplifying, the equation looks like this:

$$4r^2 + 2r - 2 = 0$$

Factoring gives you:

$$2(2r - 1)(r + 1) = 0$$

So the roots are:

$$r_1 = ½$$

and

$$r_2 = -1$$

This means that the general solution to the original equation is

$$y = c_1 x^{½} + c_2 x^{-1}$$

TIP

Note how this technique is similar to the one in the situation where the differential equation you're dealing with has constant coefficients (as in Chapter 5) and you assume a solution of the form $y = e^{rx}$. Here, there are powers of $x^2$ and $x$ already appearing in the coefficients of the differential equation, so you assume that the solution is of the form $y = x^r$.

## Real and equal roots

Sometimes the two roots of $r^2 + (\alpha - 1)\, r + \beta = 0$ are equal. For example, take a look at this differential equation:

$$x^2 \frac{d^2y}{dx^2} + 3x \frac{dy}{dx} + 2y = 0$$

Substituting $y = x^r$ into this equation gives you:

$$[r\,(r - 1) + 3\,r + 2]\, x^r = 0$$

or:

$$r\,(r - 1) + 3\,r + 2 = 0$$

After some simplifying, this becomes:

$$r^2 + 2\,r + 2 = 0$$

which in turn becomes:

$$(r + 1)(r + 1) = 0$$

So the roots here are –1 and –1. How do you handle a case like this? Clearly $y_1 = x^{-1}$, but what's $y_2$?

As you find out in Chapter 5, when you have a second order differential equation with constant coefficients, you try a solution of the following form:

$$y = e^{rx}$$

After you substitute this solution into the differential equation, you may find that the two roots, $r_1$ and $r_2$, are equal. In that case, you end up with these two solutions:

$$y_1 = e^{r_1 x}$$

and:

$$y_2 = x\, e^{r_1 x}$$

TIP

Is there an analog for Euler equations? You bet. If $r_1 = r_2$, then:

$$y_1 = x^{r_1}$$

And in that case:

$$y_2 = \ln(x)\, x^{r_1}$$

REMEMBER

So the general solution of Euler equations with equal roots of the characteristic equation is:

$$y = c_1 x^{r_1} + c_2 \ln(x)\, x^{r_1}$$

And that's it. The solution to your original equation is:

$$y = c_1 x^{-1} + c_2 \ln(x)\, x^{-1}$$

## Complex roots

In this section, you take a look at the case where the roots of a Euler equation are complex and of the form $a \pm ib$. In this case, $a$ and $b$ are constants, and $i$ is the square root of –1. Your solutions look like this:

$$y_1 = x^{a+ib}$$

and:

$$y_2 = x^{a-ib}$$

To deal with complex roots, you have to get into some trigonometry. Start by noting that this is true:

$$x^r = e^{r\ln(x)}$$

And note that as long as $x > 0$:

$$x^{a+ib} = e^{(a+ib)\ln(x)} \qquad x > 0$$

and this equals:

$$e^{(a+ib)\ln(x)} = e^{a\ln(x)} e^{ib\ln(x)} \qquad x > 0$$

which is:

$$e^{a\ln x} e^{ib\ln x} = x^a e^{ib\ln x} \qquad x > 0$$

Whew! At this point, you can use the following relation:

$$e^{mx} = \cos mx + i \sin mx \qquad x > 0$$

Here, $m$ is a constant. So after using the relation, your equation becomes:

$$x^a e^{ib\ln x} = x^a[\cos(b\ln(x)) + i \sin(b\ln(x))] \qquad x > 0$$

Absorbing the pesky factor of $i$ into $c_2$ gives you the general solution, which looks like this:

$$y = c_1 x^a \cos(b\ln(x)) + c_2 x^a \sin(b\ln(x)) \qquad x > 0$$

Now try putting all this trig to work with an example. Try solving this Euler equation, where all the coefficients are 1:

$$x^2 \frac{d^2y}{dx^2} + x\frac{dy}{dx} + y = 0$$

Trying a solution of the following form:

$$y = x^r$$

gives you:

$$[r(r-1) + r + 1]\, x^r = 0$$

or:

$$r(r-1) + r + 1 = 0$$

After some simplifying, this equation becomes:

$$r^2 + 1 = 0$$

Uh oh! The roots here are complex: $\pm i$. This means that $a = 0$ and $b = 1$, so you get:

$$y = c_1 \cos(\ln(x)) + c_2 \sin(\ln(x)) \qquad x > 0$$

This equation is only valid for $x > 0$. Let me guess: You're wondering whether you can do anything for $x < 0$. Well, actually, it can be shown that this is the way to handle the regions $x < 0$ and $x > 0$ (not $x = 0$):

$$y = c_1 \cos(\ln |x|) + c_2 \sin(\ln |x|) \qquad x > 0$$

## Putting it all together with a theorem

The following formal theorem summarizes the previous sections:

**If you have a Euler equation:**

$$\boldsymbol{x^2 \frac{d^2y}{dx^2} + \alpha x \frac{dy}{dx} + \beta y = 0}$$

**where α and β are real constants, then the solution is of this fundamental form in any interval that doesn't include the origin:**

$$y = x^r$$

**where you can find *r* by solving the characteristic equation:**

$$r\,(r + 1) + \alpha\, r + \beta = 0$$

**If the roots are real and distinct, then the general solution is of this form:**

$$y = c_1|x|^{r_1} + c_2|x|^{r_2}$$

**If the roots are real and equal, then the general solution is of this form:**

$$y = c_1|x|^{r_1} + c_2 \ln|x||x|^{r_1}$$

**And if the roots are complex, $a \pm ib$, then the general solution is of this form:**

$$y = c_1|x|^a\cos(b\ln|x|) + c_2|x|^a\sin(b\ln|x|)$$

# Figuring Series Solutions Near Regular Singular Points

In the following sections, you take a look at how to find the series solution near regular singular points. Have fun!

## Identifying the general solution

Take a look at the general second order differential equation that you want to solve:

$$P(x)\frac{d^2y}{dx^2} + Q(x)\frac{dy}{dx} + R(x)y = 0$$

TIP

You can assume that the solution has a regular singular point. For convenience, you can also assume that the singular point is at $x_0 = 0$. (If the singular point isn't at zero, you can make a substitution of variables, $x \to x - c$, where $c$ is a constant, so that it is).

Divide this equation by $P(x)$ to get:

$$\frac{d^2y}{dx^2} + p(x)\frac{dy}{dx} + q(x)y = 0$$

where:

$$p(x) = \frac{Q(x)}{P(x)}$$

and:

$$q(x) = \frac{R(x)}{P(x)}$$

From here, with a few tricks up your sleeve, you can decipher a general solution near singular points, as you find out in the following sections.

### Seeing the series of products

Because $x = 0$ is a regular singular point, it follows that $x\,p(x)$ and $x^2\,q(x)$ both have finite limits as $x \to 0$. This means that both products have convergent series of the following form (see Chapter 9 for an introduction to power series):

$$x\,p(x) = \sum_{n=0}^{\infty} p_n x^n$$

and:

$$x^2 q(x) = \sum_{n=0}^{\infty} q_n x^n$$

Both series converge for $|x| < a$ for some interval, $a > 0$.

### Substituting the series into the differential equation

Now multiply the original differential equation by $x^2$ to get the following equation:

$$x^2\frac{d^2y}{dx^2} + x\left[x\,p(x)\right]\frac{dy}{dx} + x^2 q(x)y = 0$$

Substituting the two series from the previous section into this equation gives you this result:

$$x^2\frac{d^2y}{dx^2}$$
$$+x\left[p_0 + p_1x + p_2x^2 + \ldots + p_nx^n + \ldots\right]\frac{dy}{dx}$$
$$+\left[q_0 + q_1x + q_2x^2 + \ldots + q_nx^n + \ldots\right]y = 0$$

### Recognizing a Euler equation

The previous equation looks a little difficult. But don't worry. Here's what to do: Try tackling problems of this kind by assuming that all coefficients except $p_0$ and $q_0$ are equal to 0. Doing so in the case of the example gives you:

$$x^2 \frac{d^2y}{dx^2} + x p_0 \frac{dy}{dx} + q_0 y = 0$$

Does this equation look familiar? It should — it's a Euler equation, like the ones I discuss earlier in this chapter. The fact that this looks like a Euler equation helps you see how to tackle the more general differential equation with a regular singular point.

Here's the key: If not all the coefficients (besides $p_0$ and $q_0$) are equal to 0, you have to assume that the solution is of this Euler-like form:

$$y = x^r \sum_{n=0}^{\infty} q_n x^n$$

This series is the same as this one:

$$y = \sum_{n=0}^{\infty} q_n x^{n+r}$$

The fundamental solution is a Euler solution, with the power series added in to take care of any non-Euler coefficients. In other words, even if your differential equation has a regular singular point, you can sometimes find a valid solution of the form of this power series in the region of the singular point.

## The basics of solving equations near singular points

Before I get to a numerical example, I want to walk you through the steps of solving an equation near singular points. The following sections show you everything you need to know. Fasten your seat belt!

### Determining the solution's form and differentiating

Say you start with a differential equation of the following form:

$$x^2 \frac{d^2y}{dx^2} - x\left[x p(x)\right]\frac{dy}{dx} + \left[x^2 q(x)\right]y = 0$$

where:

$$x\,p(x) = \sum_{n=0}^{\infty} p_n x^n$$

and:

$$x^2 q(x) = \sum_{n=0}^{\infty} q_n x^n$$

Here's the Euler equation that matches the differential equation:

$$x^2 \frac{d^2 y}{dx^2} - p_0 x \frac{dy}{dx} + q_0 y = 0$$

As you find out in the earlier section "Recognizing a Euler equation," you can assume that a solution to this equation is of the following form:

$$y = x^r \sum_{n=0}^{\infty} q_n x^n$$

This solution is the same as the following:

$$y = \sum_{n=0}^{\infty} q_n x^{n+r}$$

Differentiating this series gives you:

$$\frac{dy}{dx} = \sum_{n-0}^{\infty} a_n (r+n)\, x^{n+r-1}$$

And after differentiating again you get:

$$\frac{d^2 y}{dx^2} = \sum_{n=0}^{\infty} a_n (r+n)(r+n-1)\, x^{n+r-2}$$

### *Substituting series into the original equation*

Now it's time for the heavy-lifting. Substituting the previous three series into the original differential equation and collecting terms gives you this form of the differential equation:

$$a_0 f(r) x^r + \sum_{n=1}^{\infty} \left[ f(n+r) a_n + \sum_{m=0}^{n-1} a_m \left[ (m+r) p_{n-m} + q_{n-m} \right] \right] x^{n+r} = 0$$

where:

$$f(r) = r(r+1) + p_0 r + q_0$$

and

$$p_0 = \lim_{x \to 0} x\,p(x)$$

and

$$q_0 = \lim_{x \to 0} x^2 q(x)$$

Wow. What do you do next? Read on to find out.

### *Finding the indicial equation*

You know that in order for the terms on the left side of the previous equation to equal zero, every power of $x$ must equal zero. In particular, note that this means:

$$a_0 f(r)\, x^r = 0$$

Because $a_0$ and $x^r$ aren't zero, you know that:

$$f(r) = 0$$

When this equation is expanded, you get the following:

$$r(r + 1) + p_0 r + q_0 = 0$$

REMEMBER

This equation is the *indicial equation* for the original differential equation; in other words, it's the same characteristic equation for the roots you'd get if you solved the differential equation's corresponding Euler equation.

### *Working with the roots*

You use the two roots of the indicial equation to find the two solutions to the differential equation, $y_1$ and $y_2$.

Next, you can set the coefficients of $x^{n+r}$ to zero in the previous section's equation. Doing so gives you this relation:

$$f(n+r)a_n + \sum_{m=0}^{n-1} a_m\left[(m+r)p_{n-m} + q_{n-m}\right] = 0 \quad n \geq 1$$

which gives you the following *recurrence relation* (see Chapter 9 for details):

$$a_n = \frac{\sum_{m=0}^{n-1} a_m\left[(m+r)p_{n-m} + q_{n-m}\right]}{f(n+r)} \quad n \geq 1$$

You can then find the coefficients, $a_n$, from this recurrence relation.

There are two solutions to the original differential equation, each of which corresponds to the two roots, $r_1$ and $r_2$. Here's the first solution, $y_1$:

$$y_1 = |x^{r_1}|\left[1 + \sum_{n=1}^{\infty} a_n(r_1)x^n\right] \quad x \neq 0$$

So how do you go about finding the second solution, $y_2$? Well, how you do that depends on the two roots of the indicial equation. (See the later section "Taking a closer look at indicial equations" for details.)

## A numerical example of solving an equation near singular points

The stuff in the previous sections can be tough, so take a look at a numerical example to make it all clear in your mind.

### Uncovering the solution's form

To start, take a look at this differential equation:

$$2x^2 \frac{d^2y}{dx^2} - x\frac{dy}{dx} + (1+x)y = 0$$

This equation has a regular singular point at $x = 0$. How would you find a solution to this equation? Well, after you give it some thought, it looks a lot like this Euler equation:

$$2x^2 \frac{d^2y}{dx^2} - x\frac{dy}{dx} + y = 0$$

So you can try a solution of the following form:

$$y = \sum_{n=0}^{\infty} a_n x^{n+r}$$

Differentiating this solution gives you:

$$\frac{dy}{dx} = \sum_{n=0}^{\infty} a_n (r+n) x^{n+r-1}$$

After differentiating again, you get:

$$\frac{d^2y}{dx^2} = \sum_{n=0}^{\infty} a_n (r+n)(r+n-1) x^{n+r-2}$$

### Using series as substitutes in the original equation

Substituting the previous three series into your original differential equation gives you this impressive result:

$$2\sum_{n=0}^{\infty} a_n(r+n)(r+n-1)\,x^{n+r-2}$$

$$-\sum_{n=0}^{\infty} a_n(r+n)\,x^{n+r}$$

$$+\sum_{n=0}^{\infty} a_n x^{n+r}$$

$$+\sum_{n=0}^{\infty} a_n x^{n+r+1} = 0$$

TIP

This equation can also be written as:

$$a_0\left[2r(r-1) - r + 1\right]x^r$$

$$+\sum_{n=1}^{\infty}\left(\left[2(r+n)(r+n-1) - (r+n) + 1\right]a_n + a_{n-1}\right)x^{r+n} = 0$$

Wow. Can you decipher this? Keep reading to find out!

### *Unearthing the indicial equation*

Because the coefficients of all powers of $x$ in the previous equation must be zero for the whole equation to be zero, you can set the coefficient of $x^r$ to be zero, like so:

$$2r(r-1) - r + 1 = 0$$

Multiplying this out gives you the following equation:

$$2r^2 - 3r + 1 = 0$$

TIP

This turns out to be the same characteristic equation you would have received from the Euler equation that's closest to the equation that you're trying to solve, namely:

$$2x^2 \frac{d^2y}{dx^2} - x\,\frac{dy}{dx} + y = 0$$

You can factor $2r^2 - 3r + 1 = 0$ into the following indicial equation:

$$(r-1)(2r-1) = 0$$

As you can see, the roots of this equation are:

$$r_1 = 1$$

and

$$r_2 = \tfrac{1}{2}$$

These two roots are given a big name: *exponents at the singularity.* They're given this name because they describe the behavior of the solution near the singular point, $x = 0$. You put these roots to work in the next two sections.

Getting back to the original differential equation in impressive series form, you now set the coefficients of $x^{r+n}$ equal to zero to get this:

$$[2(r+n)(r+n-1) - (r+n) + 1]a_n + a_{n-1} = 0$$

You can rewrite this equation as:

$$a_n = \frac{-a_{n-1}}{2(r+n)^2 - 3(r+n) + 1}$$

Or, with some regrouping you get:

$$a_n = \frac{-a_{n-1}}{[(r+n)-1][2(r+n)-1]}$$

For each root of the initial equation, $r_1$ and $r_2$, you can use this recurrence relation to find the coefficients $a_n$. Keep reading to find out what to do next!

### Applying the first root

Plugging the first root, $r_1 = 1$, into the recurrence relation gives you:

$$a_n = \frac{-a_{n-1}}{n(2n+1)}$$

So when you plug in 1 for $a$, you get this:

$$a_1 = \frac{-a_0}{3}$$

And when you plug 2 in for $a$ you get this:

$$a_2 = \frac{-a_1}{(5)(2)}$$

Substituting $a_1$ into this equation gives you:

$$a_2 = \frac{-a_0}{(5)(2)(3)}$$

Now here's the next coefficient:

$$a_3 = \frac{-a_2}{(7)(3)}$$

Substituting $a_2$ into this equation gives you:

$$a_3 = \frac{-a_0}{(7)(3)(5)(2)(3)}$$

TIP

You can rewrite this equation as:

$$a_3 = \frac{-a_0}{(3)(5)(7)(1)(2)(3)}$$

In general, you get this relation:

$$a_n = \frac{(-1)^n a_0}{\left[(3)(5)\ldots(2n+1)\right]n!}$$

which means that $y_1$ of the general solution is:

$$y_1 = x\left[1 + \sum_{n=1}^{\infty} \frac{(-1)^n x^n}{\left[(3)(5)\ldots(2n+1)\right]n!}\right] \quad x > 0$$

TIP

Here's the million-dollar question: Does this series converge? As you may expect, you can use the ratio test to find out (see Chapter 9 for details):

$$\lim_{n \to \infty} \left| \frac{a_{n+1} x^{n+1}}{a_n x^n} \right|$$

which works out to be:

$$\lim_{n \to \infty} \left| \frac{x}{(2n+3)(n+1)} \right|$$

And as $n \to \infty$, this limit $\to 0$, so the series expansion is valid for all values of $x$.

### Plugging in the second root

Now how about $r_2 = ½$? To work with the second root, turn back to the following recurrence relation:

$$a_n = \frac{-a_{n-1}}{\left[(r+n)-1\right]\left[2(r+n)-1\right]}$$

Plugging in $r = ½$ gives you:

$$a_n = \frac{-a_{n-1}}{\left[n - \frac{1}{2}\right]\left[2n\right]}$$

which equals:

$$a_n = \frac{-a_{n-1}}{(2n-1)n}$$

As you may know, $a_0$ is arbitrary and is set to match the initial conditions of the problem. So you start by finding $a_1$:

$$a_1 = \frac{-a_0}{(1)(1)}$$

So:

$$a_1 = -a_0$$

Now for $a_2$:

$$a_2 = \frac{-a_1}{(3)(2)}$$

Substituting $a_1$ into this equation gives you the following result:

$$a_2 = \frac{a_0}{(1)(3)(1)(2)}$$

And for $a_3$:

$$a_3 = \frac{-a_2}{(3)(5)}$$

Substituting $a_2$ into this equation gives you:

$$a_3 = \frac{-a_0}{(1)(2)(3)(1)(3)(5)}$$

And for $a_4$:

$$a_4 = \frac{-a_3}{(4)(7)}$$

Substituting $a_3$ into this equation gives you:

$$a_4 = \frac{-a_0}{(1)(2)(3)(4)(1)(3)(5)(7)}$$

So, in general:

$$a_n = \frac{(-1)^n a_0}{n!\,(1)(3)(5)(7)\ldots(2n-1)}$$

You can give the second solution, $y_2$, like this:

$$y_2 = x^{1/2}\left[1 + \sum_{n=1}^{\infty} \frac{(-1)^n x^n}{\left[(1)(3)(5)\ldots(2n-1)\,n!\right]}\right] \qquad x > 0$$

How about the radius of convergence? To find out, you can use the ratio test as you did in the previous section:

$$\lim_{n \to \infty} \left| \frac{a_{n+1} x^{n+1}}{a_n x^n} \right|$$

After some simplification, this equation can be rewritten as:

$$\lim_{n \to \infty} \left| \frac{x}{(2n-1)n} \right|$$

Congratulations! Now you know that the series converges for all $x$.

## Taking a closer look at indicial equations

In the numerical example in the previous section, there were two real and distinct roots to the indicial equation (the equation you would get if the differential equation were an Euler equation). So you got two distinct solutions, $y_1$ and $y_2$. The general solution is:

$$y = c_1 y_1 + c_2 y_2$$

What happens if the indicial equation's roots are the same — that is, real and equal? In that case, the first solution is of the following form:

$$y_1 = \sum_{n=0}^{\infty} a_n x^{n+r}$$

The second solution usually involves a logarithmic term. If the roots of the indicial equation are complex, on the other hand, the solution usually involves sines and cosines.

In the following sections, I examine a bit more closely the several types of roots you can have in indicial equations.

### Distinct roots that don't differ by a positive integer

If $r_2 \neq r_1$, and $r_1 - r_2$ isn't a positive integer, the second solution, $y_2$, is given by the following:

$$y_2 = |x^{r_2}| \left[ 1 + \sum_{n=1}^{\infty} a_n(r_2) x^n \right] \qquad x \neq 0$$

Because these two series:

$$1+\sum_{n=1}^{\infty} a_n(r_1)\,x^n$$

and:

$$1+\sum_{n=1}^{\infty} a_n(r_2)\,x^n$$

are analytic at $x = 0$. The behavior of the solutions at $x = 0$, where there's a regular singular point, is entirely due to these terms:

$$|x^{r_1}|$$

and

$$|x^{r_2}|$$

That's why $r_1$ and $r_2$ are called the exponents at the singularity, as I explain in the earlier section "Finding the indicial equation" — they determine what happens at the singular point.

### Equal roots

What if the roots $r_1$ and $r_2$ are equal? In that case, $y_2$ takes the following form:

$$y_2 = y_1 \ln|x| + |x^{r_1}| \sum_{n=1}^{\infty} b_n(r_1)\,x^n \qquad x > 0$$

You have to calculate the coefficients, $b_n$, as usual: You substitute this equation into the differential equation, collect terms, set the coefficients of $x$ equal to zero, and get a recurrence relationship working.

### Roots that differ by a positive integer

If the roots of the indicial equation differ by a positive integer, $r_1 - r_2 = N$, things get complex pretty quickly. And you can't use a solution, $y_2$, of this form:

$$y_2 = |x^{r_2}|\left[1+\sum_{n=1}^{\infty} a_n(r_2)\,x^n\right] \qquad x \neq 0$$

Why? Well, because if $r_1 - r_2 = N$, you would start overlapping with the $y_1$ solution:

$$y_1 = |x^{r_1}|\left[1+\sum_{n=1}^{\infty} a_n(r_1)\,x^n\right] \qquad x \neq 0$$

So what does the second solution look like when $r_1 - r_2 = N$? Here's what it turns out to be:

$$y_2 = a\, y_1 \ln|x| + |x^{r_2}|\left[1 + \sum_{n=1}^{\infty} c_n(r_2)\, x^n\right]$$

Here's what the constant $a$ looks like (which may be zero):

$$a = \lim_{r \to r_2} (r - r_2)\, a_N(r)$$

Here, $a_N$ is the $N$th coefficient , where $r_1 - r_2 = N$, and

$$c_n(r_2) = \frac{d}{dr}\left[(r - r_2)\, a_n(r)\right]$$

# Chapter 11

# Working with Laplace Transforms

### *In This Chapter*

- Looking at the elements of a Laplace transform
- Determining whether a Laplace transform converges
- Figuring some Laplace transforms
- Using Laplace transforms to solve differential equations
- Getting a grip on factoring, convolutions, and step functions

This chapter is all about a powerful new tool for handling particularly tough differential equations. This tool is called a *Laplace transform,* which is a type of integral transform; you use it to change a differential equation into something simpler, solve the simpler equation, and then *invert* the transform to recover the solution to your original differential equation. Cool, huh? Read on for all the details.

## Breaking Down a Typical Laplace Transform

To get to know Laplace transforms, start by taking a look at what a general integral transform looks like:

$$F(s) = \int_{\alpha}^{\beta} K(s,t) f(t)\, dt$$

In this case, $f(t)$ is the function that you're taking an integral transform of, and $F(s)$ is the transformed function. The limits of integration, $\alpha$ and $\beta$, can be anything you choose, but the most common limits are $-\infty$ to $+\infty$. And here's the key to the transform: $K(s, t)$ is called the *kernel* of the transform, and you choose your own kernel. The idea is that choosing your own kernel gives you a chance to simplify your differential equation more easily.

## Who was Laplace?

Pierre-Simon Laplace (March 23, 1749–March 5, 1827) was a French mathematician and astronomer. One of his most important contributions was his five-volume work, *Mécanique Céleste* (*Celestial Mechanics*), in which he laid the groundwork of modern mathematical astronomy. This work was used by Sir Isaac Newton. Laplace was responsible for what is now called Laplace's equation and for the Laplace transform (which is important to mathematical physics). The Laplacian operator, which is a differential operator that's central to many areas of physics, also is named after Laplace.

TIP

When you restrict yourself to differential equations with constant coefficients, as you do in this chapter, a useful kernel is $e^{-st}$. Why? Because when you differentiate this kernel with respect to *t*, you end up with powers of *s*, which you can equate to the constant coefficients. Here's what this process looks like:

$$F(s) = \int_0^{\infty} e^{-st} f(t)\, dt$$

Note that besides using the kernel $e^{-st}$, the limits of integration in the previous transform are from 0 to ∞ because negative values of *t* would make the integral diverge.

The symbol for Laplace transforms is $\mathcal{L}\left\{f(t)\right\}$, which is the Laplace transform of *f*(*t*):

$$\mathcal{L}\left\{f(t)\right\} = F(s) \int_0^{\infty} e^{-st} f(t)\, dt$$

# Deciding Whether a Laplace Transform Converges

WARNING!

A potential difficulty exists with Laplace transforms: Sometimes an integral won't converge. Frequently, integrals with infinite ranges don't converge, and if the Laplace transform of a function won't converge, it won't be of any use to you in solving differential equations.

Take the following equation, for example: $f(t) = e^t$. The Laplace transform for this equation is:

$$F(s) = \int_0^\infty e^{-st} e^t \, dt$$

So here's what the new and improved transform looks like:

$$F(s) = \int_0^\infty e^{-s} \, dt$$

Taking the $e^{-s}$ factor outside the integral leaves you with this equation:

$$F(s) = e^{-s} \int_0^\infty dt$$

This equation is all fine and good, but integrating it gives you:

$$F(s) = t e^{-s} \Big|_0^\infty$$

And for finite values of $s$, this equation tends toward infinity instead of converging, which is obviously a problem.

So how do you know if a Laplace transform exists in finite form? It's time for a theorem. But first, I start by defining what it means for the function $f(t)$ to be *piecewise continuous*. The function $f(t)$ is piecewise continuous if both of the following are true:

- $f(t)$ is continuous on each subinterval $t_{i-1} < t < t_i$, where $i$ stands for the number of the interval
- $f(t)$ stays finite as the endpoints of each subinterval are approached (from inside the subinterval)

Now that you're armed with that definition, here's the theorem that helps you determine whether a Laplace transform exists for your particular function:

**If $f(t)$ is piecewise continuous in the interval $0 \leq t \leq \alpha$ for any positive $\alpha$, and $| f(t) | \leq Ce^{at}$ where $t \geq K$, where $C$, $a$, and $K$ are real and positive, then the Laplace transform $F(x) = \mathcal{L}\{f(t)\}$ exists for $s > a$.**

# Calculating Basic Laplace Transforms

In this section, you can take a look at how to calculate some basic Laplace transforms. Enjoy!

## The transform of 1

In this section, you start off by calculating just about the easiest Laplace transform there is — the Laplace transform of 1. Here's what that transform looks like:

$$\mathcal{L}\{1\} = \int_0^\infty e^{-st}(1)\,dt$$

Integration gives you:

$$\mathcal{L}\{1\} = \int_0^\infty e^{-st}(1)\,dt = \frac{1}{s} \qquad s > 0$$

So the Laplace transform of 1 remains finite for all $s > 0$, and it depends on the value you choose for $s$.

## The transform of $e^{at}$

Now move on to solving the transform of $e^{at}$: $f(t) = e^{at}$. Here's what the Laplace transform looks like:

$$\mathcal{L}\{e^{at}\} = \int_0^\infty e^{-st}e^{at}\,dt$$

With a little simplifying, this equation becomes:

$$\mathcal{L}\{e^{at}\} = \int_0^\infty e^{-(s-a)t}\,dt$$

And this, in turn, becomes the following after integration:

$$\mathcal{L}\{e^{at}\} = \int_0^\infty e^{-(s-a)t}\,dt = \frac{1}{s-a} \qquad s > a$$

Again, this result depends on the value you choose for $s$.

## The transform of sin at

The Laplace transforms in the previous two sections don't look so bad. How about now trying your hand at the Laplace transform of some trig functions, such as sin $at$? Here's the Laplace transform of sin $at$, which I calculate in the following sections:

$$\mathcal{L}\{\sin at\} = \int_0^\infty \sin at\; e^{-st}\,dt$$

### Integrating by parts

So how do you go about tackling the integration to this tricky transform? Well, you have to be crafty here. The best way to solve this transform is to integrate by parts, giving you:

$$\mathcal{L}\{\sin at\} = \frac{e^{-st}\cos at}{a}\Big|_0^\infty - \frac{s}{a}\int_0^\infty \cos at\ e^{at}\,dt$$

Wow, did all that work buy you anything? Well, sure! It turns out that the first term is $1/a$ (substituting in 0 for $t$), so this breaks down to:

$$\mathcal{L}\{\sin at\} = \frac{1}{a} - \frac{s}{a}\int_0^\infty \cos at\ e^{at}\,dt$$

TIP

Here's where the clever part comes in. Note that the second term has become similar to the original integral, except that it uses cosine. Maybe if you integrate by parts again, you'll get back to an integral that uses sine. And if that's the case, maybe you can actually factor $\mathcal{L}\{\sin at\}$ onto the left side, which would leave you with some reasonable expression on the right.

Integrating the previous transform by parts again gives you:

$$\mathcal{L}\{\sin at\} = \frac{1}{a} - \frac{s^2}{a^2}\int_0^\infty \sin at\ e^{at}\,dt$$

### Simplifying the result

After going through the previous section, you may be asking yourself: "Is this just getting worse and worse?" Actually, no. Things truly are improving — note that the second term is $s^2/a^2$ multiplied by $\mathcal{L}\{\sin at\}$. So this transform becomes:

$$\mathcal{L}\{\sin at\} = \frac{1}{a} - \frac{s^2}{a^2}\mathcal{L}\{\sin at\}$$

TIP

It turns out that you have been clever after all, because you can recast the equation this way:

$$\mathcal{L}\{\sin at\} + \frac{s^2}{a^2}\mathcal{L}\{\sin at\} = \frac{1}{a}$$

Or even simpler, you get:

$$\frac{s^2+a^2}{a^2}\mathcal{L}\{\sin at\} = \frac{1}{a}$$

which becomes:

$$\mathcal{L}\{\sin at\} = \frac{a}{s^2+a^2} \qquad s > 0$$

Can that be it? Yes, that's $\mathcal{L}\{\sin at\}$. Great work!

## Consulting a handy table for some relief

As you see in the previous sections, things can become complex pretty fast when it comes to calculating Laplace transforms. Are you going to be expected to jump through these kinds of hoops each time you need a Laplace transform? If so, wouldn't it just be easier to hit yourself on your head with a brick?

Thankfully, you usually don't have to jump through those hoops very often. Instead, you can use some handy tables of Laplace transforms. After all, why reinvent the wheel? Check out Table 11-1, which is designed to save you a lot of work.

Keep in mind that the Laplace transform of *f*(*t*) is defined this way:

$$\mathcal{L}\{f(t)\} = F(s) = \int_0^{\infty} e^{-st} f(t)\,dt$$

So the Laplace transform depends on the value you choose for *s*.

**Table 11-1 Laplace Transforms of Common Functions**

| *Function* | *Laplace Transform* | *Restrictions* |
|---|---|---|
| 1 | $\frac{1}{s}$ | $s > 0$ |
| $e^{at}$ | $\frac{1}{s-a}$ | $s > a$ |
| $t^n$ | $\frac{n!}{s^{n+1}}$ | $s > 0$, $n$ an integer $> 0$ |
| cos *at* | $\frac{s}{s^2+a^2}$ | $s > 0$ |
| sin *at* | $\frac{a}{s^2+a^2}$ | $s > 0$ |
| cosh *at* | $\frac{s}{s^2-a^2}$ | $s > \lvert a \rvert$ |
| sinh *at* | $\frac{a}{s^2-a^2}$ | $s > \lvert a \rvert$ |
| $e^{at}\cos bt$ | $\frac{s-a}{(s^2-a^2)+b^2}$ | $s > a$ |
| $e^{at}\sin bt$ | $\frac{b}{(s^2-a^2)+b^2}$ | $s > a$ |
| $t^n e^{at}$ | $\frac{n!}{(s-a)^{n+1}}$ | $s > a$, $n$ an integer $> 0$ |
| $f(ct)$ | $\frac{1}{c}\mathcal{L}\{f(s/c)\}$ | $c > 0$ |
| $f^{(n)}(t)$ | $s^n\mathcal{L}\{f(t)\} - s^{n-1}f(0) - \ldots - s\,f^{(n-2)}(0) - f^{(n-1)}(0)$ | |

# Solving Differential Equations with Laplace Transforms

Now here comes the fun stuff: using Laplace transforms to solve differential equations. Take a look at this differential equation, for example:

$$y'' + 3y' + 2y = 0$$

The initial conditions for this equation are:

$$y(0) = 2$$

and

$$y'(0) = -3$$

"Wait a minute," you say. "I know how to solve this! You know that to solve this equation you just assume a solution of the following form:

$$y = ce^{at}$$

Then you plug into the original equation to get:

$$ca^2e^{at} + 3cae^{at} + 2ce^{at} = 0$$

Dividing by $ce^{at}$ gives you:

$$a^2 + 3a + 2 = 0$$

which becomes:

$$(a + 1)(a + 2) = 0$$

So $a = -1$ and $-2$, which means that the solution is:

$$y = c_1e^{-t} + c_2e^{-2t}$$

Now, matching the initial condition $y(0) = 2$ means that:

$$y(0) = c_1 + c_2 = 2$$

Take the first derivative of the solution, which gives you:

$$y' = -c_1e^{-t} + -2c_2e^{-2t}$$

So $y'(0) = -3$ gives you:

$$y'(0) = -c_1 + -2c_2 = -3$$

And solving $y(0) = c_1 + c_2 = 2$ and $y'(0) = -c_1 + -2c_2 = -3$ results in:

$$c_1 = 1$$

and:

$$c_2 = 1$$

So the general solution is:

$$y = e^{-t} + e^{-2t}$$

Excellent, you solved the problem with traditional techniques. Now try doing the same thing using Laplace transforms in the following sections. Having already solved the problem means that you can check the answer you get. And beginning with this relatively simple problem will show how to use Laplace transforms to solve differential equations.

## A few theorems to send you on your way

To start solving the example problem, you need a few more theorems. You've figured out how to do a number of individual Laplace transforms, but what about doing the Laplace transform of an equation like this one:

$$y'' + 3y' + 2y = 0$$

### The Laplace transform is a linear operator

The first theorem I want to introduce you to says that the Laplace transform is a linear operator:

**Because the Laplace transform is a linear operator, this is a true statement:**

$$\mathcal{L}\{c_1 f_1(t) + c_2 f_2(t)\} = c_1 \mathcal{L}\{f_1(t)\} + c_2 \mathcal{L}\{f_2(t)\}$$

**In other words, the Laplace transform of the sum of two terms is the sum of the Laplace transforms of those two terms.**

Taking the Laplace transform of the differential equation you want to solve gives you:

$$\mathcal{L}\{y'' + 3y' + 2y\}$$

which becomes:

$$\mathcal{L}\{y''\} + 3\mathcal{L}\{y'\} + 2\mathcal{L}\{y\}$$

So you've made some progress in solving differential equations already — as you can see, you can break things up by terms, which is a great help.

### *The Laplace transform of a first derivative*

Okay, so you already know something about the term $\mathcal{L}\left\{y\right\}$ — that's the Laplace transform of the function $y(x)$. But what about the $\mathcal{L}\left\{y'\right\}$ term? That brings me to the next theorem:

**Say that $f(t)$ is continuous and $f'(t)$ is piecewise continuous in an interval $0 \le t \le \alpha$ and there exist constants $C$, $\beta$, and $\delta$ such that $|f(t)| \le Ce^{\beta t}$ for $t \ge \delta$. In that case, $\mathcal{L}\left\{f'(t)\right\}$ exists for $s > \beta$, and:**

$$\mathcal{L}\left\{f'(t)\right\} = s\,\mathcal{L}\left\{f(t)\right\} - f(0)$$

### *The Laplace transforms of higher derivatives*

What about $\mathcal{L}\{f''(t)\}$, $\mathcal{L}\{f'''(t)\}$, all the way up to $\mathcal{L}\left\{f^{(n)}(t)\right\}$? It turns out that you can apply the previous equation over and over to find the higher derivatives. Here's the theorem that formalizes this application:

**Say that $f(t)$, $f'(t)$, $f''(t) \ldots f^{(n-1)}(t)$ are continuous and $f^{(n)}(t)$ is piecewise continuous in an interval $0 \le t \le \alpha$ and there exist constants $C$, $\beta$, and $\delta$ such that $|f(t)| \le Ce^{\beta t}$, $|f'(t)| \le Ce^{\beta t}$, $|f''(t)| \le Ce^{\beta t} \ldots |f^{(n-1)}(t)| \le Ce^{\beta t}$ for $t \ge \delta$. In that case, $\mathcal{L}\left\{f^{(n)}(t)\right\}$ exists for $s > \beta$, and:**

$$\mathcal{L}\left\{f^{(n)}(t)\right\} = s^{n}\,\mathcal{L}\left\{f(t)\right\} - s^{n-1}f(0) - \ldots - sf^{(n-2)}(0) - f^{(n-1)}(0)$$

## *Solving a second order homogeneous equation*

Alright, now that you're armed with the theorems from the previous section, you're ready to solve this differential equation in the following sections:

$$y'' + 3y' + 2y = 0$$

In general, here's how the process works:

1. **Figure out the Laplace transform of the differential equation.**
2. **Solve the equation algebraically.**
3. **Try to find the inverse transform.**

Using Laplace transforms in this process has one major advantage: It changes a differential equation to an algebraic equation. The only sticky part is finding the transforms and inverse transforms of the various terms in the differential equation (and that's what the tables of Laplace transforms are for).

You can generalize the solution technique used in the upcoming problem to the general second order differential equation $ay'' + by' + cy = f(t)$. It turns out that the Laplace transform of the solution to this differential equation is:

$$\mathcal{L}\{y(s)\} = \frac{(as+b)\,y(0) + ay'(0)}{as^2 + bs + c} + \frac{\mathcal{L}\{f(t)\}}{as^2 + bs + c}$$

Note that when you use this equation, you don't have to find the solution to the homogeneous version of the differential equation first — doing so isn't necessary when you use Laplace transforms.

### *Finding the Laplace transform of the equation's unknown solution*

Taking the Laplace transform of the differential equation (see the earlier section "The Laplace transform is a linear operator") gives you:

$$\mathcal{L}\{y''\} + 3\mathcal{L}\{y'\} + 2\mathcal{L}\{y\}$$

Then, using the theorem from the earlier sections on the Laplace transforms of derivatives, you get:

$$\mathcal{L}\{y''\} = s^2\mathcal{L}\{y\} - sy(0) - y'(0)$$

and

$$\mathcal{L}\{y'\} = s\mathcal{L}\{y\} - y(0)$$

After plugging these two derivatives into the Laplace transform of the original differential equation, you get this result:

$$s^2\mathcal{L}\{y\} - sy(0) - y'(0) + 3\left[s\mathcal{L}\{y\} - y(0)\right] + 2\mathcal{L}\{y\} = 0$$

Collecting terms gives you:

$$(s^2 + 3s + 2)\,\mathcal{L}\{y\} - (3+s)\,y(0) - y'(0) = 0$$

Now you can use the initial conditions:

$$y(0) = 2$$

and

$$y'(0) = -3$$

in the collected version of the equation's Laplace transform, which gives you:

$$(s^2 + 3s + 2)\,\mathcal{L}\{y\} - (6 + 2s) + 3 = 0$$

Or, more simply:

$$(s^2 + 3s + 2)\,\mathcal{L}\{y\} - 2s - 3 = 0$$

Moving all the terms to the correct spots gives you this result:

$$\mathcal{L}\{y\} = \frac{2s + 3}{(s^2 + 3s + 2)}$$

Factoring the denominator leaves you with this equation for the Laplace transform of the differential equation's solution:

$$\mathcal{L}\{y\} = \frac{2s + 3}{(s + 1)(s + 2)}$$

### Discovering a function to match the Laplace transform

REMEMBER

Now you have to find a function whose Laplace transform is the same as the previous solution. To do that, use the method of partial fractions, and then write the solution as:

$$\mathcal{L}\{y\} = \frac{2s + 3}{(s + 1)(s + 2)} = \frac{a}{(s + 1)} + \frac{b}{(s + 2)}$$

And now you have to figure out what $a$ and $b$ are. To do so, write the fractions as:

$$\mathcal{L}\{y\} = \frac{2s + 3}{(s + 1)(s + 2)} = \frac{a(s + 2) + b(s + 1)}{(s + 1)(s + 2)}$$

At this point, you can equate the numerators in the fractions to get:

$$2s + 3 = a\,(s + 2) + b(s + 1)$$

Because it's up to you to choose $s$, you can first set it to –1 (to make it easy on yourself), which gives you:

$$1 = a$$

And now you can set $s$ to –2 to get:

$$-1 = -b$$

or:

$$1 = b$$

Okay, so $a = 1 = b$. You have these fractions:

$$\mathcal{L}\{y\} = \frac{a}{(s+1)} + \frac{b}{(s+2)}$$

Substituting for $a$ and $b$ gives you a more detailed Laplace transform of the differential equation's solution:

$$\mathcal{L}\{y\} = \frac{1}{(s+1)} + \frac{1}{(s+2)}$$

### *Uncovering the inverse Laplace transform to get the equation's solution*

Whew. After you know what the Laplace transform of the solution looks like, it's time to find the inverse Laplace transform to find the actual solution to the differential equation.

Your calculator isn't going to have an inverse Laplace transform $\mathcal{L}^{-1}\{\ \}$ button on it (and if it does, I'd be glad to buy it from you). Your best bet, then, is to use a table, like Table 11-1 earlier in this chapter, to find the form of the transform that you're dealing with. Then you simply have to find the function whose transform that is.

From Table 11-1, you can see that the Laplace transform of $e^{at}$ is:

$$\mathcal{L}\{e^{at}\} = \frac{1}{s-a} \qquad s > a$$

This transform looks promising. You can use this information to get the inverse transform. Take a look at the first term:

$$\frac{1}{(s+1)}$$

Comparing this to the Laplace transform of $e^{at}$ tells you that in this case, $a = -1$, so the first term in the differential equation's solution is:

$$y_1 = e^{-t}$$

Now look at the second term:

$$\frac{1}{(s+2)}$$

Comparing this to the Laplace transform of $e^{at}$ tells you that $a = -2$, so the second term in the differential equation's solution is:

$$y_2 = e^{-2t}$$

And the solution to the differential equation is $y = y_1 + y_2$, which equals this final result:

$$y = y_1 + y_2 = e^{-t} + e^{-2t}$$

This solution is confirmed by your earlier solution (which you determined at the very beginning of this section). But this time, you did it with Laplace transforms. Not too bad, huh?

## Solving a second order nonhomogeneous equation

Ready for another example? How about this little gem:

$$y'' + y = -15 \sin 4t$$

where

$$y(0) = 2$$

and

$$y'(0) = 5$$

"Hmm," you say, "I think I know how to solve this one as well." Good! You must have read Chapter 6! So, you likely know that you should first take a look at the homogeneous equation:

$$y'' + y = 0$$

And because it looks like sines and cosines would work, you can assume a solution of the following form:

$$y = c_1 \sin t + c_2 \sin t$$

Now you need a particular solution. The right side has a term in sin 4*t*, so try a similar term here. Doing so would give you this form for the general solution:

$$y = c_1 \sin t + c_2 \cos t + c_3 \sin 4t$$

Plugging this into $y'' + y = -15 \sin 4t$ gives you:

$$-c_1 \sin t - c_2 \cos t - 16c_3 \sin 4t + c_1 \sin t + c_2 \cos t + c_3 \sin 4t = \sin 4t$$

Or, with some simplification you get:

$$-15\, c_3 \sin t = -15 \sin 4t$$

So $c_3 = 1$, and so far, your general solution looks like this:

$$y = c_1 \sin t + c_2 \cos t + \sin 4t$$

Now you can use the initial conditions to determine $c_1$ and $c_2$. Here's $y(0)$:

$$y(0) = c_1 \sin t + c_2 \cos t + \sin 4t = c_2 = 2$$

So $c_2 = 2$. So far, your general solution looks like this:

$$y = c_1 \sin t + 2 \cos t + \sin 4t$$

Now you can determine $c_1$ with the last initial condition, which means that you have to calculate $y'(0)$. Here's what $y'$ looks like:

$$y' = c_1 \cos t - 2 \sin t + 4 \cos 4t$$

And here's $y'(0)$:

$$y'(0) = c_1 + 4 = 5$$

So $c_1 = 1$. And the general solution is:

$$y = \sin t + 2 \cos t + \sin 4t$$

Okay, very nice. You obviously have been paying attention! Now, in the following sections you can try the same thing using Laplace transforms.

### Determining the Laplace transform

Here's the differential equation you're trying to solve with the Laplace transform, in case you forgot:

$$y'' + y = -15 \sin 4t$$

To begin, take the Laplace transform of it, using the relation from the earlier section "The Laplace transforms of higher derivatives":

$$\mathcal{L}\{y''\} = s^2 \mathcal{L}\{y\} - sy(0) - y'(0)$$

So here's your differential equation:

$$s^2 \mathcal{L}\{y\} - sy(0) - y'(0) + \mathcal{L}\{y\} = \frac{-15}{(s^2 + 16)}$$

where you've put in the Laplace transform of sin $4t$, according to Table 11-1. Here are the initial conditions for this problem:

$$y(0) = 2$$

and

$$y'(0) = 5$$

Substituting these initial conditions into the previous equation gives you this result:

$$s^2 \mathcal{L}\{y\} - 2s - 5 + \mathcal{L}\{y\} = \frac{-15}{(s^2 + 16)}$$

Or, with some simplifying you get:

$$(s^2 + 1)\mathcal{L}\{y\} - 2s - 5 = \frac{-15}{(s^2 + 16)}$$

Now, taking $2s + 5$ over to the right side gives you:

$$(s^2 + 1)\mathcal{L}\{y\} = \frac{-15}{(s^2 + 16)} + 2s + 5$$

Be careful, because you're into some heavy algebra now! After the calculations, the equation works out to be:

$$(s^2 + 1)\mathcal{L}\{y\} = \frac{-15 + (2s + 5)(s^2 + 16)}{(s^2 + 16)}$$

Next, after expanding the terms you get the following equation:

$$(s^2 + 1)\mathcal{L}\{y\} = \frac{-15 + 2s^3 + 32s + 5s^2 + 80}{(s^2 + 16)}$$

which works out to become:

$$\mathcal{L}\{y\} = \frac{2s^3 + 5s^2 + 32s + 65}{(s^2 + 1)(s^2 + 16)}$$

### *Matching a function to the Laplace transform*

Wow. How the heck do you expand the previous equation using partial fractions so you can find a function to match the transform? It's actually pretty easy! You simply break it up and assume a form like this:

$$\mathcal{L}\{y\} = \frac{as + b}{(s^2 + 1)} + \frac{cs + d}{(s^2 + 16)}$$

which is:

$$\mathcal{L}\{y\} = \frac{(as+b)(s^2+16)+(cs+d)(s^2+1)}{(s^2+1)(s^2+16)}$$

Now tackle the numerator. The first term is:

$$(as + b)(s^2 + 16)$$

Multiplying this out gives you:

$$as^3 + 16as + bs^2 + 16b$$

The second term in the numerator is:

$$(cs + d)(s^2 + 1)$$

After you multiply the term out, you get:

$$cs^3 + cs + ds^2 + \mathrm{d}$$

Adding the two terms that you multiplied out gives you this form for the numerator:

$$as^3 + 16as + bs^2 + 16b + cs^3 + cs + ds^2 + d$$

Or, after some simplifying:

$$(a + c)s^3 + (b + d)s^2 + (16a + c)s + (16b + d)$$

REMEMBER

This numerator must equal the numerator in the other equation you have for the same Laplace transform, $2s^3 + 5s^2 + 32s + 65$. So equating the two numerators gives you:

$$2s^3 + 5s^2 + 32s + 65 = (a + c)s^3 + (b + d)s^2 + (16a + c)s + (16b + d)$$

Does this really buy you anything? Well, it still looks like a cubic equation. And it still *is* a cubic equation, but the powers of *s* must be equal on the two sides of the equation, so you have these relations:

$$a + c = 2$$

$$b + d = 5$$

$$16a + c = 32$$

$$16b + d = 65$$

Solving these equations for *a*, *b*, *c*, and *d* gives you the following:

$a = 2$

$b = 1$

$c = 0$

$d = 1$

Here's your equation for the Laplace transform:

$$\mathcal{L}\{y\} = \frac{as + b}{(s^2 + 1)} + \frac{cs + d}{(s^2 + 16)}$$

Expanding this gives you:

$$\mathcal{L}\{y\} = \frac{as}{(s^2 + 1)} + \frac{b}{(s^2 + 1)} + \frac{cs}{(s^2 + 16)} + \frac{d}{(s^2 + 16)}$$

Finally, plugging in values for *a*, *b*, *c*, and *d* gives you:

$$\mathcal{L}\{y\} = \frac{2s}{(s^2 + 1)} + \frac{1}{(s^2 + 1)} + \frac{1}{(s^2 + 16)}$$

### Using the handy table to find the inverse Laplace transform

So the general solution, after you check Table 11-1 and plug in all the right numbers, is:

$$y = \sin t + 2\cos t + \sin 4t$$

As you may have noticed, this solution agrees with the one you got earlier in this chapter using the traditional method. Great work.

## Solving a higher order equation

Try out this ever-popular higher order differential equation in the land of Laplace transforms.

$$y^{(4)} - y = 0$$

where:

$y(0) = 0$

$y'(0) = 1$

$y''(0) = 0$

$y'''(0) = 0$

So now you probably want to know how to go about solving $y^{(4)} - y = 0$ using Laplace transforms. The following sections will show you how.

### *Figuring out the equation's Laplace transform*

A fourth derivative may seem dreadful, but the fact that three of the four initial conditions are zero will definitely help. Here's the Laplace transform of $y^{(4)} - y = 0$:

$$s^4 \mathcal{L}\{y\} - s^3 y(0) - s^2 y'(0) - s\, y''(0) - y'''(0) - \mathcal{L}\{y\} = 0$$

TIP

This equation looks like it has the potential of being a tough nut to crack, but substituting the initial conditions simplifies things, giving you this:

$$s^4 \mathcal{L}\{y\} - s^2 - \mathcal{L}\{y\} = 0$$

Looks a lot more manageable, doesn't it? Solving for

$$\mathcal{L}\{y\} = \frac{(as+b)(s^2+16)+(cs+d)(s^2+1)}{(s^2+1)(s^2+16)}$$ $\mathcal{L}\{y\}$ gives you:

$$\mathcal{L}\{y\} = \frac{s^2}{s^4 - 1}$$

### *Unearthing a function to match the Laplace transform*

Now the plan is to get the transform from the previous section into a form that's recognizable in Table 11-1. You can use partial fractions to do just that. Because $s^4 - 1 = (s^2 + 1)(s^2 - 1)$, you can write the previous transform in the following format:

$$\mathcal{L}\{y\} = \frac{as+b}{s^2-1} + \frac{cs+d}{s^2+1}$$

Adding these terms together gives you:

$$\mathcal{L}\{y\} = \frac{(as+b)(s^2+1)+(cs+d)(s^2-1)}{s^4-1}$$

REMEMBER

The numerator here must equal the numerator in the other equation you have for $\mathcal{L}\{y\}$. Equating the numerators gives you:

$$s^2 = (as + b)(s^2 + 1) + (cs + d)(s^2 - 1)$$

TIP

Don't forget that you can choose the value of $s$ yourself. For example, if you choose $s$ to equal 1, the previous equation simplifies to:

$$1 = 2(a + b)$$

How about if you choose $s = -1$? Then you get this result:

$$1 = 2(-a + b)$$

Adding these two equations together gives you:

$$2 = 4b$$

So:

$$b = ½$$

Substituting $b = ½$ into $1 = 2(a + b)$ gives you:

$$1 = 2(a + ½)$$

Or with some simplification:

$$1 = 1 + 2a$$

So:

$$a = 0$$

Okay, so $a = 0$ and $b = 1$. How about $c$ and $d$? You can set $s = 0$ in $2s^2 = (as + b)(s^2 + 1) + (cs + d)(s^2 - 1)$ to get:

$$0 = b - d$$

Or:

$$b = d$$

And because $b = ½$, $d = ½$ also.

Alright, that gives you $a$, $b$, and $d$. But you still have to figure out $c$. Take a look at your equation that equated the two numerators:

$$s^2 = (as + b)(s^2 + 1) + (cs + d)(s^2 - 1)$$

Note that equating the cubic terms on the two sides gives you this equation:

$$0 = as^3 + cs^3$$

But because $a = 0$, this becomes:

$$0 = cs^3$$

So $c = 0$. That gives you:

$a = 0$

$b = ½$

$c = 0$

$d = ½$

Substituting these numbers into the Laplace transform gives you this more detailed result:

$$\mathcal{L}\{y\} = \frac{\frac{1}{2}}{s^2 - 1} + \frac{\frac{1}{2}}{s^2 + 1}$$

So, as you can see, having used the initial conditions in the problem has significantly simplified the form of the Laplace transform.

### Getting the equation's inverse Laplace transform

After you have the Laplace transform, check out Table 11-1 for the inverse transform, and then plug in the correct numbers. Here's the result you should get.

$$y = \frac{1}{2}(\sinh t + \sin t)$$

# Factoring Laplace Transforms and Convolution Integrals

Sometimes, the Laplace transforms you end up with require a little extra work to solve. But no worries; you just have to be a little creative. And in the following sections, I explain how to do just that. For instance, I show you how to factor a Laplace transform into a sum of fractions and how to work with convolution integrals.

## Factoring a Laplace transform into fractions

TIP

When you face an especially funky-looking Laplace transform, factoring the denominator and the numerator often helps. For example, imagine that you end up with this Laplace transform when solving a differential equation:

$$\mathcal{L}\{y\} = \frac{s + 4}{s^2 + 4s + 8}$$

There's no entry in Table 11-1 that matches this transform. But the form of the denominator indicates that you may be able to factor it. Here's what the factoring may look like:

$$s^2 + 4s + 8 = (s + 2)^2 + 2^2$$

This equation looks somewhat more promising, don't you think? You can now convert your original transform into this new form:

$$\mathcal{L}\{y\} = \frac{s+4}{(s+2)^2 + 2^2}$$

Hey, wait a minute. This expression isn't in Table 11-1 either. But take a closer look at the numerator; you can factor it as well. For instance, you can rewrite the numerator like this:

$$s + 4 = (s + 2) + 2$$

This in turn gives you the following for the Laplace transform:

$$\mathcal{L}\{y\} = \frac{(s+2)+2}{(s+2)^2 + 2^2}$$

Or, with a little simplifying you get:

$$\mathcal{L}\{y\} = \frac{(s+2)}{(s+2)^2 + 2^2} + \frac{2}{(s+2)^2 + 2^2}$$

Doesn't this look a whole lot better? According to Table 11-1, you can now find the inverse Laplace transform, which is:

$$\mathcal{L}\{y\} = e^{-2t}\cos 2t + e^{-2t}\sin 2t$$

## Checking out convolution integrals

The example in the previous section factored a Laplace transform into a sum of fractions. But what if, instead of a sum, you end up with a product of recognizable Laplace transforms? What do you do then?

For example, what if you had this Laplace transform:

$$\mathcal{L}\{y\} = \frac{a}{s^2(s+a)^2}$$

Looking at Table 11-1, you can see that the inverse transform of this expression doesn't appear. However, if you write this Laplace transform as:

$$\mathcal{L}\{y\} = \frac{1}{s^2}\,\frac{a}{(s+a)^2}$$

then you can see that this is the product of the Laplace transforms for $t$ and sin $at$. This product brings you to an important question: Can you write the inverse Laplace transform this way:

$$\mathcal{L}\{y\} = t\sin at$$

Nope, it doesn't work like that. The inverse of the sum of Laplace transforms is the sum of the inverses of the individual Laplace transforms, but it doesn't work that way when you multiply Laplace transforms together. Instead, you have to turn to *convolution integrals,* as defined in the following theorem:

**If:**

$$\mathcal{L}\{f(t)\}\mathcal{L}\{g(t)\} = \mathcal{L}\{h(t)\}$$

**then:**

$$h(t) = \int_0^t f(t-\alpha)g(\alpha)d\alpha$$

**which also equals:**

$$h(t) \int_0^t f(\alpha)g(t-\alpha)d\alpha$$

**where $\alpha$ is a new variable. The integrals in these transforms are called convolution integrals.**

Sometimes you may see the convolution operation denoted with an asterisk (*), like this:

$$h(t) = (f * g)(t)$$

Using this new operator means that you can write the products of Laplace transforms in much the same way that you would use the multiplication operator:

$$f(t) * 0 = 0 * f(t) = 0$$

$$f(t) * g(t) = g(t) * f(t)$$

$$f(t) * (g(t) + i(t)) = f(t) * g(t) + f(t) * i(t)$$

$$(f(t) * g(t)) * i(t) = f(t) * (g(t) * i(t))$$

Armed with this theorem, you can now tackle the example problem:

$$\mathcal{L}\{y\} = \frac{1}{s^2}\frac{a}{(s+a)^2}$$

These terms are the Laplace transforms of *t* and sin *at*. So you can write the inverse Laplace transform like this:

$$y(t)\int_0^t (t-\alpha)\sin(at)\,d\alpha$$

which equals:

$$y(t)=\frac{t}{a}-\frac{\sin at}{a^2}$$

# Surveying Step Functions

In the world of Laplace transforms, there's an interesting function called the *step function,* which looks like this: $u_a(t)$. In the following sections, I break down the elements of the step function, and I show how it relates to Laplace transforms.

## Defining the step function

The function's value is defined as 0 up to a certain point, *a*, but then as 1 thereafter. Here's the definition of the step function in equation form:

$$u_a(t)=\begin{cases}0 & t<a\\ 1 & t\geq a\end{cases}$$

As an example, if $a = 0$, you have that $u_0(t)$ equals:

$$u_0(t)=\begin{cases}0 & t<0\\ 1 & t\geq 0\end{cases}$$

Note that the subscript on the *u* indicates the point at which the step function "steps."

And if $a = 1$, you have that $u_1(t)$ equals:

$$u_1(t)=\begin{cases}0 & t<1\\ 1 & t\geq 1\end{cases}$$

You can see what $u_a(t)$ looks like graphically in Figure 11-1.

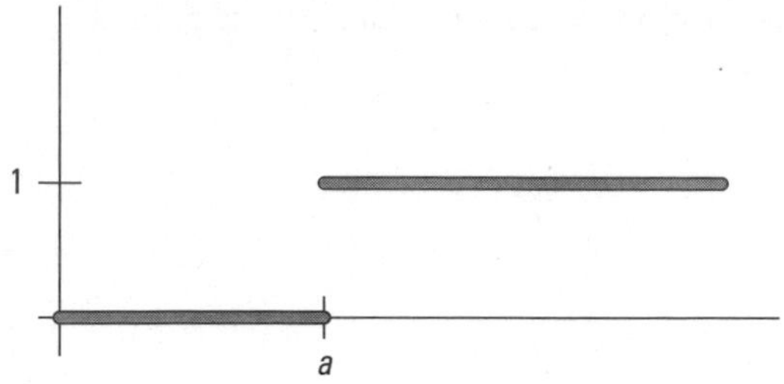


**Figure 11-1:** The step function in graph form.

## Figuring the Laplace transform of the step function

Now that you're familiar with the step function, go a bit further and take the Laplace transform of the step function:

$$\mathcal{L}\{u_a(t)\} = \int_0^{\infty} e^{-st} u_a(t)\, dt$$

Substituting the step function into this transform gives you the following result:

$$\mathcal{L}\{u_a(t)\} = \int_a^{\infty} e^{-st}\, dt$$

Note that the lower limit of the integral has become $a$. Why? Because the integral is zero up to that point. Now this equation becomes:

$$\mathcal{L}\{u_a(t)\} = \frac{e^{-as}}{s} \qquad s > 0$$

And that brings me to a new theorem:

**If $a$ is a positive constant, then:**

$$\mathcal{L}\{u_a(t) f(t-a)\} = e^{-as} \mathcal{L}\{f(t)\}$$

This theorem can help you with Laplace transforms that have exponentials in them. To see how, try finding the inverse Laplace transform of this equation:

$$\mathcal{L}\{y\} = \frac{1 - e^{-s}}{s^2}$$

Begin by breaking it into the following:

$$\mathcal{L}\{y\} = \frac{1}{s^2} - \frac{e^{-s}}{s^2}$$

Using the theorem, you can find the inverse Laplace transform:

$$y(t) = t - u_1(t)(t - 1)$$

That is, for $t < 1$:

$$y(t) = t$$

and for $t \geq 1$:

$$y(t) = 1$$

So the step function helped you find a Laplace transform. Pretty cool, right?

# Chapter 12

# Tackling Systems of First Order Linear Differential Equations

### *In This Chapter*

- Brushing up on matrix basics
- Handling matrix operations
- Checking out linear independence, eigenvalues, and eigenvectors
- Working out homogeneous and nonhomogeneous systems

This chapter is all about systems of differential equations, where a *system* is a set of linear differential equations that share some common variables. In this chapter, I show you how to apply many of the techniques used to handle systems of standard algebraic equations to solving systems of differential equations.

Being able to handle systems of differential equations is useful in two cases:

- When you want to reduce the order of a differential equation to a system of first order differential equations
- When you have a problem that consists of interdependent differential equations — such as when you have an electrical circuit with linked loops in it that share the current (don't worry; I won't delve into such science here)

This chapter is heavy in its use of *matrices,* which are usually used to solve systems of linear equations. After you brush up on the fundamentals of matrices, you find out about some important concepts, such as linear independence, eigenvalues, and eigenvectors. I wrap up the chapter by explaining how to solve both homogeneous and nonhomogeneous systems.

# Introducing the Basics of Matrices

Take a look at the following system of three simultaneous equations:

$$x + y + z = 6$$
$$x - y - z = -4$$
$$x + y - z = 0$$

How do you solve this system of equations for $x$, $y$, and $z$? A fair bit of algebra is involved here, so to keep things straight, you can use a matrix. You're most likely already familiar with matrices, but in the following sections, I go over the fundamentals just in case you need to refresh your memory.

## Setting up a matrix

REMEMBER

Simply put, a *matrix* is a set of numbers arranged into columns and rows, like this:

$$\begin{pmatrix} a_1 & a_2 & a_3 \\ b_1 & b_2 & b_3 \\ c_1 & c_2 & c_3 \end{pmatrix}$$

When creating a matrix, you place the coefficients of each equation in a row. For instance, $a_1$, $a_2$, and $a_3$ are from one equation; $b_1$, $b_2$, and $b_3$ are from a second equation; and $c_1$, $c_2$, and $c_3$ are from a third equation. Similarly, the first coefficient of any equation is in the first column; the second coefficient of any equation is in the second column; and the third coefficient of any equation is in the third column. (I'm sure you get the idea!) A matrix can contain as many rows and columns as you need.

Putting numbers into rows and columns like this helps you keep track of them. (Neatniks rejoice!) For example, you can represent the coefficients in these equations:

$$x + y + z = 6$$
$$x - y - z = -4$$
$$x + y - z = 0$$

like this:

$$\begin{pmatrix} 1 & 1 & 1 \\ 1 & -1 & -1 \\ 1 & 1 & -1 \end{pmatrix}$$

And to keep track of the constants on the right side of these equations, you can create what's called an *augmented matrix.* With this type of matrix, a vertical line keeps the constants on the right sides of the equations separate from the coefficients on the left sides of the equations, like this:

$$\begin{pmatrix} 1 & 1 & 1 \\ 1 & -1 & -1 \\ 1 & 1 & -1 \end{pmatrix} \begin{pmatrix} 6 \\ -4 \\ 0 \end{pmatrix}$$

## Working through the algebra

After you set up a matrix, it's easy to keep track of the algebraic manipulations involved in solving the system of simultaneous equations. Here, you're going to work on the individual rows.

You can see how this works using the matrix you set up in the previous section. Remember that your goal is to solve for $x$, $y$, and $z$. First, add the second row to the first and place the result in the first row, making sure that you leave the second row intact. Doing all this gives you:

$$\begin{pmatrix} 2 & 0 & 0 \\ 1 & 1 & 1 \\ 1 & 1 & 1 \end{pmatrix} \begin{pmatrix} 2 \\ -4 \\ 0 \end{pmatrix}$$

Now divide the first row by 2 to simplify:

$$\begin{pmatrix} 1 & 0 & 0 \\ 1 & -1 & -1 \\ 1 & 1 & -1 \end{pmatrix} \begin{pmatrix} 1 \\ -4 \\ 0 \end{pmatrix}$$

Next, add –1 times the first row to the second row and place the result in the second row:

$$\begin{pmatrix} 1 & 0 & 0 \\ 0 & -1 & -1 \\ 1 & 1 & 1 \end{pmatrix} \begin{pmatrix} 1 \\ -5 \\ 0 \end{pmatrix}$$

Then multiply the second row by –1:

$$\begin{pmatrix} 1 & 0 & 0 \\ 0 & 1 & 1 \\ 1 & 1 & -1 \end{pmatrix} \begin{pmatrix} 1 \\ 5 \\ 0 \end{pmatrix}$$

Add –1 times the first row to the third row and place the result in the third row, like so:

$$\begin{pmatrix} 1 & 0 & 0 \\ 0 & 1 & 1 \\ 0 & 1 & -1 \end{pmatrix} \begin{pmatrix} 1 \\ 5 \\ -1 \end{pmatrix}$$

Now add –1 times the second row to the third, placing the result in the third row:

$$\begin{pmatrix} 1 & 0 & 0 \\ 0 & 1 & 1 \\ 0 & 0 & -2 \end{pmatrix} \begin{pmatrix} 1 \\ 5 \\ -6 \end{pmatrix}$$

Finally, you divide the third row by –2:

$$\begin{pmatrix} 1 & 0 & 0 \\ 0 & 1 & 1 \\ 0 & 0 & 1 \end{pmatrix} \begin{pmatrix} 1 \\ 5 \\ 3 \end{pmatrix}$$

TIP

Note that you now have a *triangular matrix* on the bottom left; in other words, the bottom left triangle's elements all equal zero. This triangle simplifies matters considerably, because this new augmented matrix now represents this system:

$$x = 1$$

$$y + z = 5$$

$$z = 3$$

And from the latter two equations, finding the solution for $y$ is pretty trivial: $y = 2$. Now you've solved for all three variables. Nice work.

## Examining matrices

REMEMBER

Did you know that you can name matrices? Yup, that's right. In matrix terms, a name is denoted in **bold.** For instance, you can name the following matrix, which contains both real and complex elements, **A** (imaginative, I know):

$$\mathbf{A} = \begin{pmatrix} 1 & 2+2i \\ 3+3i & 4 \end{pmatrix}$$

The *transpose* of a matrix swaps its rows and columns; in other words, the rows become the columns, and the columns become the rows. Here's what the transpose of **A** (called $\mathbf{A}^T$) looks like:

$$\mathbf{A}^T = \begin{pmatrix} 1 & 3+3i \\ 2+2i & 4 \end{pmatrix}$$

You can also define the *complex conjugate* of a matrix. To denote a complex conjugate, you simply add a line above the name of a matrix. The complex conjugate of a matrix is simply the complex conjugate element by element of the matrix. Here's the complex conjugate of **A**, where you flip the sign of the imaginary part:

$$\overline{\mathbf{A}} = \begin{pmatrix} 1 & 2-2i \\ 3-3i & 4 \end{pmatrix}$$

The *adjoint* of a matrix is the transpose of the complex conjugate of the matrix. You denote an adjoint by adding an asterisk (*) to the name of the matrix. Here's what $\mathbf{A}^*$ looks like:

$$\mathbf{A}^* = \begin{pmatrix} 1 & 3-3i \\ 2-2i & 4 \end{pmatrix}$$

# Mastering Matrix Operations

After you're familiar with the basics of matrices, you can take the next step and start working with a variety of matrix operations, such as addition, subtraction, multiplication, and other fun stuff. I explain what you need to know in the following sections.

## Equality

REMEMBER

Two matrices are considered equal if every element in the first matrix is equal to the corresponding element in the other matrix. For instance, if:

$$\mathbf{A} = \begin{pmatrix} 1 & 2 \\ 3 & 4 \end{pmatrix}$$

And:

$$\mathbf{B} = \begin{pmatrix} 1 & 2 \\ 3 & 4 \end{pmatrix}$$

then **A** = **B**. Easy enough, right?

## Addition

If you need to, you can add two matrices together. Doing so involves adding the elements at corresponding positions in the two matrices. For example, if:

$$\mathbf{A} = \begin{pmatrix} 1 & 2 \\ 3 & 4 \end{pmatrix} \text{ and } \mathbf{B} = \begin{pmatrix} 5 & 6 \\ 7 & 8 \end{pmatrix}$$

then:

$$\mathbf{A} + \mathbf{B} = \begin{pmatrix} 1 & 2 \\ 3 & 4 \end{pmatrix} + \begin{pmatrix} 5 & 6 \\ 7 & 8 \end{pmatrix} = \begin{pmatrix} 6 & 8 \\ 10 & 12 \end{pmatrix}$$

And just so you know: **A** + **B** = **B** + **A**.

## Subtraction

Like matrix addition, matrix subtraction works as you'd expect: element by element. For example, take a look:

$$\mathbf{A} - \mathbf{B} = \begin{pmatrix} 1 & 2 \\ 3 & 4 \end{pmatrix} - \begin{pmatrix} 5 & 6 \\ 7 & 8 \end{pmatrix} = \begin{pmatrix} -4 & -4 \\ -4 & -4 \end{pmatrix}$$

Note that **A – B** doesn't equal **B – A.** In fact, as you'd expect, **A – B** = –(**B – A**), as you can see here:

$$\mathbf{B} - \mathbf{A} = \begin{pmatrix} 5 & 6 \\ 7 & 8 \end{pmatrix} - \begin{pmatrix} 1 & 2 \\ 3 & 4 \end{pmatrix} = \begin{pmatrix} 4 & 4 \\ 4 & 4 \end{pmatrix}$$

## Multiplication of a matrix and a number

In some cases, you need to multiply a matrix by a number; to do so, you multiply each element in the matrix by that number. For example, 4**A** looks like this:

$$4\mathbf{A} = 4\begin{pmatrix} 1 & 2 \\ 3 & 4 \end{pmatrix} = \begin{pmatrix} 4 & 8 \\ 12 & 16 \end{pmatrix}$$

## Multiplication of two matrices

How about multiplying two matrices together, such as **A** and **B**? Unfortunately, this is a little more involved than just adding them. It turns out that **AB** is defined when the number of columns in **A** is the same as the number of rows in **B**. That is, if **A** is an $l \times m$ matrix (that's row $\times$ column notation, so **A** has $l$ rows and $m$ columns) and **B** is an $m \times n$ matrix, the product **AB** exists — and the product is an $l \times n$ matrix.

If **AB** = **C**, here's how it works: the $(i, j)$ (that's row, column) element of **C** is found by multiplying each element of the $i$th row of **A** by the matching element in the $j$th column of **B**. Then when you add the products you get:

$$C_{ij} = \sum_{k=1}^{m} A_{ik} B_{kj} = A_{i1} B_{1j} + A_{i2} B_{2j} + A_{i3} B_{3j} + \ldots + A_{im} B_{mj}$$

Now put in the numbers:

$$\mathbf{AB} = \begin{pmatrix} 1 & 2 \\ 3 & 4 \end{pmatrix} \begin{pmatrix} 5 & 6 \\ 7 & 8 \end{pmatrix}$$

Applying the rules of multiplying matrices, you get:

$$\mathbf{AB} = \begin{pmatrix} 1 & 2 \\ 3 & 4 \end{pmatrix} \begin{pmatrix} 5 & 6 \\ 7 & 8 \end{pmatrix} = \begin{pmatrix} 5+14 & 6+16 \\ 15+28 & 18+32 \end{pmatrix} = \begin{pmatrix} 19 & 22 \\ 43 & 50 \end{pmatrix}$$

One thing to note is that in general, $\mathbf{AB} \neq \mathbf{BA}$. Using the same example, here's what **BA** looks like:

$$\mathbf{BA} = \begin{pmatrix} 5 & 6 \\ 7 & 8 \end{pmatrix} \begin{pmatrix} 1 & 2 \\ 3 & 4 \end{pmatrix} = \begin{pmatrix} 5+18 & 10+24 \\ 7+24 & 14+32 \end{pmatrix} = \begin{pmatrix} 23 & 34 \\ 31 & 46 \end{pmatrix}$$

## Multiplication of a matrix and a vector

You'll often see matrices that only have one set of rows or one set of columns. These matrices are called *vectors*. For example, here's a vector with one column:

$$\begin{pmatrix} 1 \\ 2 \\ 3 \end{pmatrix}$$

Here's a vector with one row:

$$\begin{pmatrix} 1 & 2 & 3 \end{pmatrix}$$

Sometimes you'll see vectors that are used to hold variables, as in this one, which I'll call **x**:

$$\mathbf{x} = \begin{pmatrix} x \\ y \\ z \end{pmatrix}$$

With vectors, you can express a set of simultaneous equations like this:

$$\begin{pmatrix} 1 & 2 \\ 3 & 4 \\ 5 & 6 \end{pmatrix} \begin{pmatrix} x \\ y \\ z \end{pmatrix} = \begin{pmatrix} 3 \\ 4 \\ 5 \end{pmatrix}$$

This set of simultaneous equations can be written as:

$$\mathbf{Ax} = \mathbf{b}$$

## Identity

The *identity matrix,* which is labeled **I**, holds 1s along its upper-left to lower-right diagonal, and all 0s otherwise. Here's a 2 × 2 identity matrix (with 2 rows and 2 columns):

$$\mathbf{I} = \begin{pmatrix} 1 & 0 \\ 0 & 1 \end{pmatrix}$$

And here's a 3 × 3 identity matrix (with 3 rows and 3 columns):

$$\mathbf{I} = \begin{pmatrix} 1 & 0 & 0 \\ 0 & 1 & 0 \\ 0 & 0 & 1 \end{pmatrix}$$

Multiplying any matrix, **A**, by the identity matrix gives you **A** again. (I explain how to multiply two matrices together earlier in this chapter.) For example, take a look at the multiplication of a matrix called **A** and the 3 × 3 identity matrix:

$$\mathbf{AI} = \begin{pmatrix} 1 & 2 & 3 \\ 4 & 5 & 6 \\ 7 & 8 & 9 \end{pmatrix} \begin{pmatrix} 1 & 0 & 0 \\ 0 & 1 & 0 \\ 0 & 0 & 1 \end{pmatrix} = \begin{pmatrix} 1 & 2 & 3 \\ 4 & 5 & 6 \\ 7 & 8 & 9 \end{pmatrix}$$

## The inverse of a matrix

If you have simultaneous equations such as these:

$$x + y + z = 6$$

$$x - y - z = -4$$

$$x + y - z = 0$$

you can write a system in matrix form like this:

$$\begin{pmatrix} 1 & 1 & 1 \\ 1 & -1 & -1 \\ 1 & 1 & -1 \end{pmatrix} \begin{pmatrix} x \\ y \\ z \end{pmatrix} = \begin{pmatrix} 6 \\ -4 \\ 0 \end{pmatrix}$$

You can further simplify this so that it looks like:

$$\mathbf{Ax} = \mathbf{b}$$

where:

$$\mathbf{A} = \begin{pmatrix} 1 & 1 & 1 \\ 1 & -1 & -1 \\ 1 & 1 & -1 \end{pmatrix}$$

$$\mathbf{x} = \begin{pmatrix} x \\ y \\ z \end{pmatrix}$$

$$\mathbf{b} = \begin{pmatrix} 6 \\ -4 \\ 0 \end{pmatrix}$$

Wouldn't it be nice if you could solve for **x** by some way dividing **Ax** = **b** by **A** to get the following:

$$\mathbf{x} = \mathbf{A}^{-1}\mathbf{b}$$

Doing so *would* give you the solutions, $x$, $y$, and $z$, to your simultaneous equations. So can you find $\mathbf{A}^{-1}$? Yes, you sure can! This is called the *inverse* of **A** (and $\mathbf{A}^{-1}\,\mathbf{A} = \mathbf{I}$, where **I** is the identity matrix). So how do you find the inverse of a matrix? You can choose from several ways, but the easiest is the one that I present in the following sections.

Generally, to find the inverse of a matrix **A**, apply the steps to **A** in order to reduce it to the identity matrix **I** (that is, 1s along the upper-left to lower-right diagonal and 0s otherwise). Then apply those same steps, in the same order, to the identity matrix. You should end up with $\mathbf{A}^{-1}$. To wrap up, simply solve for **x** by multiplying $\mathbf{A}^{-1}$ and **b**.

Before I can explain how to find an inverse, I need to cover an important concept that's helpful when you're working with inverses: determinants. Read on for details.

### Finding the determinant of a matrix

An important quantity when it comes to matrices is the *determinant.* The determinant reduces a matrix to a single significant number; whether that number is zero is important for inverses as well as other calculations. For example, when checking to see whether the solutions you get to a system of linear differential equations are linearly independent (and form a general solution), you need to use determinants. (I discuss linear independence in more detail later in this chapter.)

So how do you find a matrix's determinant? Here's an example that can show you how. Say you have this 2 × 2 matrix, called **A**:

$$\mathbf{A} = \begin{pmatrix} a & b \\ c & d \end{pmatrix}$$

The determinant of **A**, written as det(**A**), for a 2 × 2 matrix is defined this way:

$$\det(\mathbf{A}) = ad - cb$$

For example, here's a 2 × 2 matrix with some specific numbers:

$$\mathbf{A} = \begin{pmatrix} 1 & 2 \\ 3 & 4 \end{pmatrix}$$

What's its determinant? Just plug the numbers into $ad - cb$, like so:

$$\det(\mathbf{A}) = (1)(4) - (3)(2) = -2$$

How about a 3 × 3 determinant? Say you have this 3 × 3 matrix **A**:

$$\mathbf{A} = \begin{pmatrix} a & b & c \\ d & e & f \\ g & h & i \end{pmatrix}$$

Be careful here! Determinants rapidly become more complex beyond 2 × 2. Here's what the determinant of the 3 × 3 matrix is:

$$\det(\mathbf{A}) = aei - afh - bdi + cdh + bfg - ceg$$

### Finding the inverse of a matrix

To find the inverse of a matrix **A**, apply the steps you need to **A** to reduce it to the identity matrix, **I** (that is, 1s along the upper-left to lower right diagonal and 0s otherwise). Then apply those same steps, in the same order, to an identity matrix, and you'll end up with $\mathbf{A}^{-1}$.

Take a look at an example. For this system of equations:

$$x + y + z = 6$$

$$x - y - z = -4$$

$$x + y - z = 0$$

matrix **A** looks like this:

$$\mathbf{A} = \begin{pmatrix} 1 & 1 & 1 \\ 1 & -1 & -1 \\ 1 & 1 & -1 \end{pmatrix}$$

Find $\mathbf{A}^{-1}$ by first reducing **A** to **I**. To do so, add –1 times the first row to the third row and place the result in the third row:

$$\begin{pmatrix} 1 & 1 & 1 \\ 1 & -1 & -1 \\ 0 & 0 & -2 \end{pmatrix}$$

Now divide the third row by –2 to simplify:

$$\begin{pmatrix} 1 & 1 & 1 \\ 1 & -1 & -1 \\ 0 & 0 & 1 \end{pmatrix}$$

Add the second row to the first row and place the result in the first row:

$$\begin{pmatrix} 2 & 0 & 0 \\ 1 & -1 & -1 \\ 0 & 0 & 1 \end{pmatrix}$$

Next, divide the first row by 2 to simplify:

$$\begin{pmatrix} 1 & 0 & 0 \\ 1 & 1 & 1 \\ 0 & 0 & 1 \end{pmatrix}$$

You're looking good. Now add –1 times the first row to the second row and place the result in the second row:

$$\begin{pmatrix} 1 & 0 & 0 \\ 0 & -1 & -1 \\ 0 & 0 & 1 \end{pmatrix}$$

Almost there. Add the third row to the second row and place the result in the second row:

$$\begin{pmatrix} 1 & 0 & 0 \\ 0 & -1 & 0 \\ 0 & 0 & 1 \end{pmatrix}$$

Finally, multiply the second row by –1 to get that final 1:

$$\begin{pmatrix} 1 & 0 & 0 \\ 0 & 1 & 0 \\ 0 & 0 & 1 \end{pmatrix}$$

And there you have it — the identity matrix.

After you reduce a system to the identity matrix, you have to apply the same sequence of steps to the identity matrix to find $\mathbf{A}^{-1}$. Here's the $3 \times 3$ identity matrix:

$$\mathbf{I} = \begin{pmatrix} 1 & 0 & 0 \\ 0 & 1 & 0 \\ 0 & 0 & 1 \end{pmatrix}$$

Now follow the same steps that you followed earlier in this section. First, add –1 times the first row to the third row and place the result in the third row:

$$\begin{pmatrix} 1 & 0 & 0 \\ 0 & 1 & 0 \\ -1 & 0 & 1 \end{pmatrix}$$

Divide the third row by –2:

$$\begin{pmatrix} 1 & 0 & 0 \\ 0 & 1 & 0 \\ 1/2 & 0 & -1/2 \end{pmatrix}$$

Next, add the second row to the first row and place the result in the first row:

$$\begin{pmatrix} 1 & 1 & 0 \\ 0 & 1 & 0 \\ 1/2 & 0 & -1/2 \end{pmatrix}$$

Divide the first row by 2:

$$\begin{pmatrix} 1/2 & 1/2 & 0 \\ 0 & 1 & 0 \\ 1/2 & 0 & -1/2 \end{pmatrix}$$

And add –1 times the first row to the second row and place the result in the second row:

$$\begin{pmatrix} \frac{1}{2} & \frac{1}{2} & 0 \\ -\frac{1}{2} & \frac{1}{2} & 0 \\ \frac{1}{2} & 0 & -\frac{1}{2} \end{pmatrix}$$

Now add the third row to the second row, placing the result in the second row:

$$\begin{pmatrix} \frac{1}{2} & \frac{1}{2} & 0 \\ 0 & \frac{1}{2} & -\frac{1}{2} \\ \frac{1}{2} & 0 & -\frac{1}{2} \end{pmatrix}$$

Finally, multiply the second row by –1:

$$\begin{pmatrix} \frac{1}{2} & \frac{1}{2} & 0 \\ 0 & -\frac{1}{2} & \frac{1}{2} \\ \frac{1}{2} & 0 & -\frac{1}{2} \end{pmatrix}$$

And there you have it: $\mathbf{A}^{-1}$.

### *Solving for x with a little multiplication*

Now you have to find **x**, the solution to your system of equations, by using the equation $\mathbf{x} = \mathbf{A}^{-1}\mathbf{b}$. Your equation looks like this when you fill in the numbers:

$$\begin{pmatrix} x \\ y \\ z \end{pmatrix} = \begin{pmatrix} \frac{1}{2} & \frac{1}{2} & 0 \\ 0 & -\frac{1}{2} & \frac{1}{2} \\ \frac{1}{2} & 0 & -\frac{1}{2} \end{pmatrix} \begin{pmatrix} 6 \\ -4 \\ 0 \end{pmatrix}$$

Performing matrix multiplication (as I describe earlier in this chapter) gives you:

$$\begin{pmatrix} x \\ y \\ z \end{pmatrix} = \begin{pmatrix} 1 \\ 2 \\ 3 \end{pmatrix}$$

Voila! It worked. You found the solution to your simultaneous equations by finding the inverse of a matrix.

WARNING!

Does solving for **x** always work? The answer is a resounding no. Why? Sometimes matrices can't be inverted (in which case they're called *singular*, or *non-invertable*, matrices).

How can you tell if a matrix is singular before attempting to find its inverse? You can check to see whether its determinant is zero. Matrices whose determinants are zero are singular matrices without an inverse; matrices whose determinants aren't zero do have an inverse. In the original matrix in my example, the determinant is:

$$\det(\mathbf{A}) = aei - afh - bdi + cdh + bfg - ceg$$

# Having Fun with Eigenvectors 'n' Things

You're now ready for more insight into matrices using eigenvectors and eigenvalues. What are those? Read on to find out more.

## Linear independence

Determining whether a set of vectors is *linearly dependent* can be quite important. It's especially important when you're solving systems of linear differential equations to see whether you have *linearly independent* solutions. You need to get a linearly independent set of solutions to make sure you have the complete solution.

Say, for example, that you have a set of vectors, $\mathbf{x}_1$ to $\mathbf{x}_k$. They're said to be linearly dependent if there are constants, $c_1$ to $c_k$, such that:

$$c_1\mathbf{x}_1 + c_2\mathbf{x}_2 + \ldots c_k\mathbf{x}_k = \mathbf{0}$$

where $\mathbf{0}$ is a vector whose elements are all zero. In other words, $\mathbf{x}_1$ to $\mathbf{x}_k$ are considered linearly dependent if a linear relation relates them. If there are no constants, $c_1$ to $c_k$ (not all zero), such that this equation holds, the vectors, $\mathbf{x}_1$ to $\mathbf{x}_k$, are linearly independent.

How do you determine whether a set of vectors, $\mathbf{x}_1$ to $\mathbf{x}_k$, is linearly independent? You can assemble all the vectors, $\mathbf{x}_1$ to $\mathbf{x}_k$, into a single matrix and then check the determinant. If the determinant is zero, the vectors are linearly dependent. If the determinant isn't zero, the vectors are linearly independent. I walk you through an example in the following sections.

### Assembling the vectors into one matrix

Say that you have the following vectors:

$$\mathbf{x}_1 = \begin{pmatrix} 2 \\ 4 \\ -2 \end{pmatrix}$$

$$\mathbf{x}_2 = \begin{pmatrix} 2 \\ 1 \\ 3 \end{pmatrix}$$

$$\mathbf{x}_3 = \begin{pmatrix} -8 \\ 2 \\ -22 \end{pmatrix}$$

Now you can assemble these vectors into a 3 × 3 matrix called **x**:

$$\mathbf{x} = \begin{pmatrix} 2 & 2 & -8 \\ 4 & 1 & 2 \\ -2 & 3 & -22 \end{pmatrix}$$

### *Determining the determinant*

Okay, so now you have to figure out whether the determinant of **x** is zero. As discussed earlier in this chapter, the determinant of a 3 × 3 matrix **A**:

$$\mathbf{A} = \begin{pmatrix} a & b & c \\ d & e & f \\ g & h & i \end{pmatrix}$$

is the following:

$$\det(\mathbf{A}) = aei - afh - bdi + cdh + bfg - ceg$$

So the determinant of **x** in my example is:

$$(2)(1)(-22) - (2)(2)(3) - (2)(4)(-22) + (-8)(4)(3) + (2)(2)(-2) - (-8)(1)(-2)$$

which becomes:

$$-44 - 12 + 176 - 96 - 8 - 16 = 0$$

As you can see, the determinant in this case is zero. Uh-oh. That's no good.

The determinant is zero, which means that $\mathbf{x}_1$, $\mathbf{x}_2$, and $\mathbf{x}_3$ are linearly dependent. What should you do next? Remember the following:

$$c_1\mathbf{x}_1 + c_2\mathbf{x}_2 + \ldots c_k\mathbf{x}_k = 0$$

What are $c_1$, $c_2$, and $c_3$? You can figure out those constants by solving the augmented matrix. Here's your matrix:

$$\begin{pmatrix} 2 & 2 & -8 \\ 4 & 1 & 2 \\ -2 & 3 & -22 \end{pmatrix}$$

Add the first row to the third row and place the result in the third row:

$$\begin{pmatrix} 2 & 2 & -8 \\ 4 & 1 & 2 \\ 0 & 5 & -30 \end{pmatrix} \begin{pmatrix} 0 \\ 0 \\ 0 \end{pmatrix}$$

Next, divide the first row by 2 to simplify:

$$\begin{pmatrix} 1 & 1 & -4 \\ 4 & 1 & 2 \\ 0 & 5 & -30 \end{pmatrix} \begin{pmatrix} 0 \\ 0 \\ 0 \end{pmatrix}$$

Now add –4 times the first row to the second row (to get a zero in the first position of Row 2). Place the result in the second row:

$$\begin{pmatrix} 1 & 1 & -4 \\ 0 & -3 & 18 \\ 0 & 5 & -30 \end{pmatrix} \begin{pmatrix} 0 \\ 0 \\ 0 \end{pmatrix}$$

Divide the second row by –3 to simplify:

$$\begin{pmatrix} 1 & 1 & -4 \\ 0 & 1 & -6 \\ 0 & 5 & -30 \end{pmatrix} \begin{pmatrix} 0 \\ 0 \\ 0 \end{pmatrix}$$

Add –5 times the second row to the third, and then place the result in the third row:

$$\begin{pmatrix} 1 & 1 & -4 \\ 0 & 1 & -6 \\ 0 & 0 & 0 \end{pmatrix} \begin{pmatrix} 0 \\ 0 \\ 0 \end{pmatrix}$$

This matrix corresponds to a system of these two equations:

$$c_1 + c_2 - 4c_3 = 0$$

$$c_2 - 6c_3 = 0$$

TIP

You can choose one of these constants arbitrarily. So, just for kicks, set $c_3 = 1$. If you plug this constant into the previous two equations and do a little algebra, you'll find that $c_2 = 6$ and $c_1 = -2$.

### Putting it all together

Now it's time to assemble your constants and your vectors so you can check your work. So:

$$c_1\begin{pmatrix}2\\4\\-2\end{pmatrix}+c_2\begin{pmatrix}2\\1\\3\end{pmatrix}+c_3\begin{pmatrix}-8\\2\\-22\end{pmatrix}$$

becomes:

$$-2\begin{pmatrix}2\\4\\-2\end{pmatrix}+6\begin{pmatrix}2\\1\\3\end{pmatrix}+\begin{pmatrix}-8\\2\\-22\end{pmatrix}=\begin{pmatrix}0\\0\\0\end{pmatrix}$$

As you can see, you've determined $c_1$, $c_2$, and $c_3$ correctly.

## Eigenvalues and eigenvectors

Take a look at this vector equation:

$$\mathbf{Ax} = \mathbf{y}$$

REMEMBER

When facing problems like this, it sometimes helps to consider **A** in a form where a linear transformation converts **x** into **y**. That is, you want to look for solutions of the following form:

$$\mathbf{Ax} = \lambda\mathbf{x}$$

In this case, λ stands for a constant. You can rewrite this equation as:

$$(\mathbf{A} - \lambda\mathbf{I})\mathbf{x} = 0$$

This equation has a solution only if $(\mathbf{A} - \lambda\mathbf{I})^{-1}$ exists. In other words, you can be assured of a solution if:

$$\det(\mathbf{A} - \lambda\mathbf{I}) \neq 0$$

REMEMBER

Any values of λ that satisfy $(\mathbf{A} - \lambda\mathbf{I})\,\mathbf{x} = 0$ are called *eigenvalues* of the original equation. And the solutions **x** to $(\mathbf{A} - \lambda\mathbf{I})\,\mathbf{x} = 0$ are called the *eigenvectors*. You can see how eigenvalues and eigenvectors work in the example that I go through in the following sections.

### Changing the matrix to the right form

Here's an example you can wrap your brain around: Try finding the eigenvalues and eigenvectors of the following matrix:

$$\mathbf{A} = \begin{pmatrix} -1 & -1 \\ 2 & -4 \end{pmatrix}$$

First, convert the matrix into the $\mathbf{A} - \lambda\mathbf{I}$ form. Doing so gives you:

$$\mathbf{A} - \lambda\mathbf{I} = \begin{pmatrix} -1 & -\lambda & -1 \\ 2 & -4 & -\lambda \end{pmatrix}$$

### Figuring out the eigenvalues

Next, it's time to find the determinant:

$$\det(\mathbf{A} - \lambda\mathbf{I}) = (-1 - \lambda)(-4 - \lambda) + 2$$

or:

$$\det(\mathbf{A} - \lambda\mathbf{I}) = \lambda^2 + 5\lambda + 6$$

which can be factored into the following:

$$\det(\mathbf{A} - \lambda\mathbf{I}) = \lambda^2 + 5\lambda + 6 = (\lambda + 2)(\lambda + 3)$$

So, by equating this equation to 0, the eigenvalues of **A** are $\lambda_1 = -2$ and $\lambda_2 = -3$.

### Calculating the eigenvectors

How about finding the eigenvectors of **A**? To find the eigenvector corresponding to $\lambda_1$, you have to substitute $\lambda_1$ (–2) into the $\mathbf{A} - \lambda\mathbf{I}$ matrix, like so:

$$\mathbf{A} - \lambda\mathbf{I} = \begin{pmatrix} 1 & -1 \\ 2 & -2 \end{pmatrix}$$

Because $(\mathbf{A} - \lambda\mathbf{I})\mathbf{x} = 0$, you have:

$$\begin{pmatrix} 1 & -1 \\ 2 & -2 \end{pmatrix} \begin{pmatrix} x_1 \\ x_2 \end{pmatrix} = \begin{pmatrix} 0 \\ 0 \end{pmatrix}$$

And, because every row of this matrix equation must be true, you can assume that $x_1 = x_2$. So, up to an arbitrary constant, the eigenvector corresponding to $\lambda_1$ is:

$$c\begin{pmatrix} 1 \\ 1 \end{pmatrix}$$

You can also drop the arbitrary constant, and write this eigenvector as:

$$\begin{pmatrix} 1 \\ 1 \end{pmatrix}$$

How about the eigenvector corresponding to $\lambda_2$? By plugging $\lambda_2$ (–3) into the $\mathbf{A} - \lambda\mathbf{I}$ matrix, you get:

$$\mathbf{A} - \lambda\mathbf{I} = \begin{pmatrix} 2 & -1 \\ 2 & -1 \end{pmatrix}$$

Then you have the following:

$$\begin{pmatrix} 2 & -1 \\ 2 & -1 \end{pmatrix} \begin{pmatrix} x_1 \\ x_2 \end{pmatrix} = \begin{pmatrix} 0 \\ 0 \end{pmatrix}$$

So $2x_1 - x_2 = 0$ and $x_1 = x_2/2$. This means that up to an arbitrary constant, the eigenvector corresponding to $\lambda_2$ is:

$$c\begin{pmatrix} 1 \\ 2 \end{pmatrix}$$

Again, you can drop the arbitrary constant, and simply write this eigenvector as:

$$\begin{pmatrix} 1 \\ 2 \end{pmatrix}$$

# Solving Systems of First-Order Linear Homogeneous Differential Equations

You've likely discovered how to work with systems of linear equations in algebra classes. If so, you're in luck because working with systems of linear differential equations follows the same techniques. In the following sections, I give you a good look at systems of first order linear homogeneous differential equations with constant coefficients like this:

$$x_1' = x_1 + x_2$$

$$x_2' = 4x_1 + x_2$$

These differential equations are linked — that is, both contain $x_1$ and $x_2$. As such, they have to be solved together. (Flip to Chapter 5 for an introduction to working with constant coefficients in homogeneous equations.) I show you how in the following sections.

## Understanding the basics

You can write a set of equations such as $x_1' = x_1 + x_2$ and $x_2' = 4x_1 + x_2$ in this matrix form:

$$\begin{pmatrix} x_1' \\ x_2' \end{pmatrix} = \begin{pmatrix} 1 & 1 \\ 4 & 1 \end{pmatrix} \begin{pmatrix} x_1 \\ x_2 \end{pmatrix}$$

Or, shorter still, you can write the matrix like this:

$$\mathbf{x}' = \mathbf{A}\mathbf{x}$$

Here, $\mathbf{x}'$, $\mathbf{A}$, and $\mathbf{x}$ are all matrices:

$$\mathbf{x}' = \begin{pmatrix} x_1' \\ x_2' \end{pmatrix}$$

$$\mathbf{A} = \begin{pmatrix} 1 & 1 \\ 4 & 1 \end{pmatrix}$$

$$\mathbf{x} = \begin{pmatrix} x_1 \\ x_2 \end{pmatrix}$$

If $\mathbf{A}$ is a matrix of constant coefficients, you can assume a solution of the following form:

$$\mathbf{x} = \xi e^{rt}$$

Okay, you say, what's that funny squiggle ($\xi$) doing there? Actually, it's the Greek letter xi in lowercase form. It couldn't possibly stand for eigenvector, could it? You're right! You're way ahead of me.

Substituting your supposed form of solution, $\mathbf{x} = \xi e^{rt}$, into the system of differential equations, $\mathbf{x}' = \mathbf{A}\mathbf{x}$, gives you:

$$r\xi e^{rt} = \mathbf{A}\xi e^{rt}$$

where $r$ is a constant. Do you see how using matrix notation makes working with systems of differential equations easier? Now you can subtract $\mathbf{A}\xi e^{rt}$ from both sides of the equation and do a little rearranging to get:

$$(\mathbf{A} - r\mathbf{I})\xi e^{rt} = 0$$

or:

$$(\mathbf{A} - r\mathbf{I})\xi = 0$$

REMEMBER

As you may recall from the earlier section on eigenvalues and eigenvectors, this is exactly the equation that specifies the eigenvalues and eigenvectors of the matrix **A**. So the solution to the system of differential equations, $\mathbf{x}' = \mathbf{A}\mathbf{x}$, is $\mathbf{x} = \xi e^{rt}$ (provided that $r$ is an eigenvalue of **A** and $\xi$ is the associated eigenvector).

## Making your way through an example

Now take a closer look at the example I give in the previous section:

$$\begin{pmatrix} x_1' \\ x_2' \end{pmatrix} = \begin{pmatrix} 1 & 1 \\ 4 & 1 \end{pmatrix} \begin{pmatrix} x_1 \\ x_2 \end{pmatrix}$$

Because this matrix has constant coefficients, you can try a solution of the following form:

$$\mathbf{x} = \xi e^{rt}$$

I walk you through the steps of solving this system in the following sections.

### Getting the right matrix form

The next step you have to tackle is substituting $\mathbf{x} = \xi e^{rt}$ into the matrix, which gives you:

$$\begin{pmatrix} r\xi_1 e^{rt} \\ r\xi_2 e^{rt} \end{pmatrix} = \begin{pmatrix} 1 & 1 \\ 4 & 1 \end{pmatrix} \begin{pmatrix} \xi_1 e^{rt} \\ \xi_2 e^{rt} \end{pmatrix}$$

By subtracting the left side from both sides of the equation, you can rewrite this as:

$$\begin{pmatrix} 0 \\ 0 \end{pmatrix} = \begin{pmatrix} 1-r & 1 \\ 4 & 1-r \end{pmatrix} \begin{pmatrix} \xi_1 e^{rt} \\ \xi_2 e^{rt} \end{pmatrix}$$

Dividing by $e^{rt}$ gives you:

$$\begin{pmatrix} 0 \\ 0 \end{pmatrix} = \begin{pmatrix} 1-r & 1 \\ 4 & 1-r \end{pmatrix} \begin{pmatrix} \xi_1 \\ \xi_2 \end{pmatrix}$$

### Finding the eigenvalues

This system of linear equations has a solution only if the determinant of the $2 \times 2$ matrix is zero (see the earlier section "Linear independence" for details). So:

$$\det\begin{pmatrix} 1-r & 1 \\ 4 & 1-r \end{pmatrix} = 0$$

TIP

After expanding, your equation looks like this:

$$(1 - r)(1 - r) - 4 = 0$$

which becomes:

$$r^2 - 2r + 1 - 4 = 0$$

Or, with some simplification:

$$r^2 - 2r - 3 = 0$$

Finally, you can factor this into:

$$(r - 3)(r + 1) = 0$$

So the eigenvalues of your $2 \times 2$ matrix are:

$$r_1 = 3$$

and

$$r_2 = -1$$

### Coming up with the eigenvectors

After you find the eigenvalues, you need to find the two eigenvectors. Plugging the first eigenvalue from the previous section, $r_1 = 3$, into the matrix yields this result:

$$\begin{pmatrix} 0 \\ 0 \end{pmatrix} = \begin{pmatrix} -2 & 1 \\ 4 & -2 \end{pmatrix} \begin{pmatrix} \xi_1 \\ \xi_2 \end{pmatrix}$$

With some simplification you get:

$$-2\xi_1 + \xi_2 = 0$$

and:

$$4\xi_1 - 2\xi_2 = 0$$

These equations are the same up to a factor of –2. So you get:

$$2\xi_1 = \xi_2$$

The first eigenvector is (up to, as usual, an arbitrary nonzero constant that doesn't matter):

$$\begin{pmatrix} \xi_1 \\ \xi_2 \end{pmatrix} = \begin{pmatrix} 1 \\ 2 \end{pmatrix}$$

How about the second eigenvector? It corresponds to the eigenvalue $r_2 = -1$. So plugging that into the matrix gives you:

$$\begin{pmatrix} 0 \\ 0 \end{pmatrix} = \begin{pmatrix} 2 & 1 \\ 4 & 2 \end{pmatrix} \begin{pmatrix} \xi_1 \\ \xi_2 \end{pmatrix}$$

After simplification you get:

$$2\xi_1 + \xi_2 = 0$$

and:

$$4\xi_1 + 2\xi_2 = 0$$

These equations give you the same information as with the first eigenvector:

$$2\xi_1 = -\xi_2$$

So the second eigenvector becomes:

$$\begin{pmatrix} \xi_1 \\ \xi_2 \end{pmatrix} = \begin{pmatrix} -1 \\ 2 \end{pmatrix}$$

### Summing up the solution

The first solution to the original system of differential equations, based on the first eigenvector and eigenvalue, is:

$$\mathbf{x}_1 = \begin{pmatrix} 1 \\ 2 \end{pmatrix} e^{3t}$$

The second solution, based on the second eigenvector and eigenvalue, is:

$$\mathbf{x}_2 = \begin{pmatrix} -1 \\ 2 \end{pmatrix} e^{-t}$$

The general solution to this system is a linear combination of these two solutions ($c_1\mathbf{x}_1 + c_2\mathbf{x}_2$), which in this example looks like:

$$\mathbf{x} = c_1 \begin{pmatrix} 1 \\ 2 \end{pmatrix} e^{3t} + c_2 \begin{pmatrix} -1 \\ 2 \end{pmatrix} e^{-t}$$

The solution can also be written as:

$$\begin{pmatrix} x_1 \\ x_2 \end{pmatrix} = c_1 \begin{pmatrix} 1 \\ 2 \end{pmatrix} e^{3t} + c_2 \begin{pmatrix} -1 \\ 2 \end{pmatrix} e^{-t}$$

TIP

By splitting up this equation, the solutions to the system of differential equations become:

$$x_1 = c_1\, e^{3t} - c_2\, e^{-t}$$

and

$$x_2 = 2c_1\, e^{3t} + 2c_2\, e^{-t}$$

That example wasn't so bad, was it?

# Solving Systems of First Order Linear Nonhomogeneous Equations

Now I want you to turn your attention to systems of first order linear nonhomogeneous differential equations. I show you how to solve this system of homogeneous differential equations in the previous section:

$$\begin{pmatrix} x_1' \\ x_2' \end{pmatrix} = \begin{pmatrix} 1 & 1 \\ 4 & 1 \end{pmatrix} \begin{pmatrix} x_1 \\ x_2 \end{pmatrix}$$

But what if the situation was changed to this:

$$\begin{pmatrix} x_1' \\ x_2' \end{pmatrix} = \begin{pmatrix} 1 & 1 \\ 4 & 1 \end{pmatrix} \begin{pmatrix} x_1 \\ x_2 \end{pmatrix} + \begin{pmatrix} -2e^{-t} \\ 3t \end{pmatrix}$$

REMEMBER

The solution to the nonhomogeneous system is the general solution of the homogeneous version of the system (which you already have from the previous section) plus a particular solution.

In the following sections, I show you how to assume the form of the solution and determine the missing coefficients — all in a way entirely analogous to the method of undetermined coefficients in linear nonhomogeneous differential equations (see Chapter 6 for an introduction to this method).

## Assuming the correct form of the particular solution

So how can you find a particular solution to a nonhomogeneous system? Easy: By using the method of undetermined coefficients after it has been generalized to work with matrices. To do that, rewrite the system like this:

$$\begin{pmatrix} x_1' \\ x_2' \end{pmatrix} = \begin{pmatrix} 1 & 1 \\ 4 & 1 \end{pmatrix} \begin{pmatrix} x_1 \\ x_2 \end{pmatrix} + \begin{pmatrix} -2e^{-t} \\ 0 \end{pmatrix} + \begin{pmatrix} 0 \\ 3t \end{pmatrix}$$

It's clear that the last two terms in this system involve terms in $e^{-t}$ and $t$, so can you assume a solution of the following form?

$$\mathbf{x} = \mathbf{a}e^{-t} + \mathbf{b}t + \mathbf{c}$$

Well, not so fast. As you may recall from the previous section, the eigenvalue of the general solution to the homogenous equation here was –1. So there's already a $e^{-t}$ term in the general solution.

What do you do in cases like this? You do just what you would do for a single differential equation — you have to add a term in $te^{-t}$. That makes your assumed solution look like this:

$$\mathbf{x} = \mathbf{a}te^{-t} + \mathbf{b}e^{-t} + \mathbf{c}t + \mathbf{d}$$

Here's what your system of differential equations looks like currently:

$$\mathbf{x} = \mathbf{A}\,x + \begin{pmatrix} -2e^{-t} \\ 0 \end{pmatrix} + \begin{pmatrix} 0 \\ 3t \end{pmatrix}$$

Plugging $\mathbf{x} = \mathbf{a}te^{-t} + \mathbf{b}e^{-t} + \mathbf{c}t + \mathbf{d}$ into this system gives you:

$$\mathbf{a}e^{-t} - \mathbf{a}te^{-t} - \mathbf{b}e^{-t} + \mathbf{c} = \mathbf{Aa}te^{-t} + \mathbf{Ab}e^{-t} + \mathbf{Ac}t + \mathbf{Ad} + \begin{pmatrix} -2e^{-t} \\ 0 \end{pmatrix} + \begin{pmatrix} 0 \\ 3t \end{pmatrix}$$

And equating coefficients on the two sides gives you these results:

$$\mathbf{Aa} = -\mathbf{a}$$

$$\mathbf{Ab} = \mathbf{a} - \mathbf{b} - \begin{pmatrix} -2 \\ 0 \end{pmatrix}$$

$$\mathbf{Ac} = -\begin{pmatrix} 0 \\ 3 \end{pmatrix}$$

$$\mathbf{Ad} = \mathbf{c}$$

## Crunching the numbers

After assuming the correct form of your solution, you're ready to do the math. In the following sections, I explain how to calculate all your missing coefficients. Remember that **A** is the following:

$$\mathbf{A} = \begin{pmatrix} 1 & 1 \\ 4 & 1 \end{pmatrix}$$

### Finding a

When you start calculating the missing coefficients, why not start with finding **a**? In this case, $\mathbf{Aa} = -\mathbf{a}$ becomes:

$$\begin{pmatrix} 1 & 1 \\ 4 & 1 \end{pmatrix}\begin{pmatrix} a_1 \\ a_2 \end{pmatrix} = -\begin{pmatrix} a_1 \\ a_2 \end{pmatrix}$$

Or, in simpler terms:

$$a_1 + a_2 = -a_1$$

and:

$$4a_1 + a_2 = -a_2$$

These can be rewritten even more simply as:

$$2a_1 = -a_2$$

and:

$$4a_1 = -2a_2$$

These equations differ by a factor of 2. Because you're only interested in a single particular solution, go with $a_1 = -1$ and $a_2 = 2$, which gives you:

$$\mathbf{a} = \begin{pmatrix} -1 \\ 2 \end{pmatrix}$$

### Finding b

Now find **b**. You already know the following:

$$\mathbf{Ab} = \mathbf{a} - \mathbf{b} - \begin{pmatrix} -2 \\ 0 \end{pmatrix}$$

So after plugging in your solution for **a**, you get:

$$\begin{pmatrix} b_1 + b_2 \\ 4b_1 + b_2 \end{pmatrix} = \begin{pmatrix} -1 \\ 2 \end{pmatrix} - \begin{pmatrix} b_1 \\ b_2 \end{pmatrix} - \begin{pmatrix} -2 \\ 0 \end{pmatrix}$$

Or, more simply:

$$2b_1 + b_2 = 1$$

and

$$4b_1 + 2b_2 = 2$$

A solution to these equations (which only differ by a factor of 2) is:

$$\mathbf{b} = \begin{pmatrix} 1 \\ -1 \end{pmatrix}$$

### *Finding c*

Ready to find *c*? You already know the following:

$$\mathbf{Ac} = -\begin{pmatrix} 0 \\ 3 \end{pmatrix}$$

So in other words, you get:

$$\begin{pmatrix} 1 & 1 \\ 4 & 1 \end{pmatrix} \begin{pmatrix} c_1 \\ c_2 \end{pmatrix} = -\begin{pmatrix} 0 \\ 3 \end{pmatrix}$$

Here's what you get after simplifying:

$$c_1 + c_2 = 0$$

and:

$$4c_1 + c_2 = -3$$

The solution to these equations, after a little number crunching, is:

$$\mathbf{c} = \begin{pmatrix} -1 \\ 1 \end{pmatrix}$$

### Finding d

Finally you just have to find **d**. You know that **Ad** = **c**. In other words, by plugging in your solution for **c** you get:

$$d_1 + d_2 = -1$$

and:

$$4d_1 + d_2 = 1$$

So the solution to this pair of equations is:

$$\mathbf{d} = \begin{pmatrix} 2/3 \\ -5/3 \end{pmatrix}$$

## Winding up your work

After going through the previous sections and assuming a form of the solution and crunching the numbers, you just have to put the work together to get a particular solution to your original system. That particular solution is:

$$x = \begin{pmatrix} x_1 \\ x_2 \end{pmatrix} = \begin{pmatrix} -1 \\ 2 \end{pmatrix} te^{-t} + \begin{pmatrix} 1 \\ -1 \end{pmatrix} e^{-t} + \begin{pmatrix} -1 \\ 1 \end{pmatrix} t + \begin{pmatrix} 2/3 \\ -5/3 \end{pmatrix}$$

# Chapter 13

# Discovering Three Fail-Proof Numerical Methods

### *In This Chapter*

- Brushing up on the Euler method
- Taking it to the next level with the improved Euler method
- Getting great results with the Runge-Kutta method

A group of elite computer scientists shuffles into your office with a problem: "We're not comfortable writing software that solves differential equations. Isn't there an easier, more computer-friendly way of handling differential equations?"

"Well," you say, "you can use the method of undetermined coefficients and then solve for . . ."

"Solve for?" the group asks. "You mean using variables? No, no — we want something *numeric.*"

"Ah," you tell them. "You want methods like Euler's method and the Runge-Kutta method." You show them some programming code.

"That's it!" the group cries, grabbing your code and running off.

"My bill," you call after them, "will be in the mail."

"Better use e-mail," cries a junior member, disappearing around the corner.

Differential equations can stump even the best and brightest, but I have a secret weapon for you: numerical methods. This chapter is all about the computer-based methods that you can use to solve differential equations when everything else fails. Do remember that you'll get numbers out of these techniques, not elegant, finished formulas. But sometimes, numbers are just what you want, as is often the case in engineering.

In this chapter, I use the Java programming language, so if you want to follow along, you need to install Java on your computer. It's free when you visit `java.sun.com`. When you're at the Web site, simply click the Java SE link and download and install the Java Development Kit, or JDK.

# Number Crunching with Euler's Method

Euler's method, which I introduce in Chapter 4, allows you to handle differential equations in a numeric way. In the following sections, I explain the basics of the method, and then I show you how to enter code into your computer so you can see the method in action.

## The fundamentals of the method

Take a look at this differential equation:

$$\frac{dy}{dx} = f(x, y)$$

The standard Euler's method notes that you may not have the actual function that represents the solution to your differential equation. However, when you're armed with the preceding equation, you do have the *slope* of that curve everywhere. That is, the rate of change of the curve is the derivative. (See Chapter 1 for a refresher on derivatives.)

Say that you have a point, $(x_0, y_0)$, that's on the solution curve. Because of the preceding equation, you know that the slope of the solution curve at that point is $f(x_0, y_0)$. Suppose that you also want to find the numeric solution at a point, $(x, y)$, a short distance, $h$, away. Here's how Euler's method says that you may find $y$:

$$y = y_0 + \Delta y$$

In other words, $y$ is equal to $y$ at an initial point, plus the change in $y$.

Because the slope, $m$, is defined as $\Delta y/\Delta x$ (a change in $y$ divided by a change in $x$), this equation also can be written like this:

$$y = y_0 + m\,\Delta x$$

And because $\Delta x = x - x_0$, this equation is also:

$$y = y_0 + m\,(x - x_0)$$

Here's the key: The slope $m$ is equal to the derivative at $(x_0, y_0)$, and because of your original differential equation, you know that $m = f(x_0, y_0)$. So you have:

$$y = y_0 + f(x_0, y_0)\,(x - x_0)$$

Now convert $(x - x_0)$ to the symbol that it usually goes by when you discuss Euler's method: $h$, which gives you this equation:

$$y_1 = y_0 + f(x_0, y_0)\,h$$

This equation can be generalized to any point, $(x_n, y_n)$, along the solution curve like this:

$$y_{n+1} = y_n + f(x_n, y_n)\,h$$

And there you have it — you've discovered the *recurrence relation,* which ties one term to the next, for Euler's equation. (Flip to Chapter 9 for an introduction to recurrence relations.)

## Using code to see the method in action

Now try testing the basics from the previous section, using this differential equation:

$$\frac{dy}{dx} = x$$

where $y(0) = 0$

I'll spare you the details of solving this equation with traditional methods (but you can head to Chapter 4 to find out how to do so). Without further ado, here's the exact solution:

$$y = \frac{x^2}{2}$$

Because you know the exact solution, you can check the accuracy of Euler's method in Java code. The following sections show you how.

### Typing in the code

In Chapter 4, I develop a short Java program, e.java, to display the results of Euler's method versus the exact solution. The code starts at $(x_0, y_0) = (0, 0)$, which you know is on the solution curve because of the initial condition, $y(0) = 0$. The code calculates 100 steps, using a step size of $h = 0.1$.

If you want to follow along, start by using the Java compiler, javac.exe, to compile e.java:

```
C:\>javac e.java
```

If javac.exe isn't in your computer's path, you have to specify that path this way to run javac.exe:

```
C:\>C:\jdk\bin\javac e.java
```

After compiling e.java, you should get a new file, e.class. Now you can execute the compiled code, e.class, using java.exe like this:

```
C:\>java e
```

Here's the code that you're going to modify in this chapter (note that the sections of the code that you have to change when you're solving your own differential equations are given in **bold**):

```
public class e
{

  double x0 = 0.0;
  double y0 = 0.0;
  double h = 0.1;
  double n = 100;

  public e()
  {
  }

  public double f(double x, double y)
  {
    return x;
  }

  public double exact(double x, double y)
  {
    return x * x / 2;
  }

  public static void main(String [] argv)
  {
    e de = new e();
    de.calculate();
  }

  public void calculate()
  {
```

```
    double x = x0;
    double y = y0;
    double k;

    System.out.println("x\t\tEuler\t\tExact");

    for (int i = 1; i < n; i++){
      k = f(x, y);
      y = y + h * k;
      x = x + h;
      System.out.println(round(x) + "\t\t" + round(y) +
           "\t\t" + round(exact(x, 0)));
    }

  }

  public double round(double val)
  {
    double divider = 100;
    val = val * divider;
    double temp = Math.round(val);
    return (double)temp / divider;
  }
}
```

### Surveying the results

After being run, the code will display the current *x* value, the Euler approximation of the solution at that value, and the exact solution, like this:

```
C:\>java e
x               Euler           Exact
0.1             0.0             0.01
0.2             0.01            0.02
0.3             0.03            0.05
0.4             0.06            0.08
0.5             0.1             0.13
0.6             0.15            0.18
0.7             0.21            0.24
0.8             0.28            0.32
0.9             0.36            0.4
1.0             0.45            0.5
1.1             0.55            0.6
1.2             0.66            0.72
1.3             0.78            0.85
1.4             0.91            0.98
1.5             1.05            1.13
1.6             1.2             1.28
1.7             1.36            1.45
1.8             1.53            1.62
```

```
1.9          1.71         1.81
2.0          1.9          2.0
2.1          2.1          2.21
2.2          2.31         2.42
2.3          2.53         2.65
2.4          2.76         2.88
2.5          3.0          3.13
2.6          3.25         3.38
2.7          3.51         3.65
2.8          3.78         3.92
2.9          4.06         4.21
3.0          4.35         4.5
3.1          4.65         4.81
3.2          4.96         5.12
3.3          5.28         5.45
3.4          5.61         5.78
3.5          5.95         6.13
3.6          6.3          6.48
3.7          6.66         6.85
3.8          7.03         7.22
3.9          7.41         7.61
4.0          7.8          8.0
4.1          8.2          8.41
4.2          8.61         8.82
4.3          9.03         9.25
4.4          9.46         9.68
4.5          9.9          10.13
4.6          10.35        10.58
4.7          10.81        11.04
4.8          11.28        11.52
4.9          11.76        12.0
5.0          12.25        12.5
5.1          12.75        13.0
5.2          13.26        13.52
5.3          13.78        14.04
5.4          14.31        14.58
5.5          14.85        15.12
5.6          15.4         15.68
5.7          15.96        16.24
5.8          16.53        16.82
5.9          17.11        17.4
6.0          17.7         18.0
6.1          18.3         18.6
6.2          18.91        19.22
6.3          19.53        19.84
6.4          20.16        20.48
6.5          20.8         21.12
6.6          21.45        21.78
6.7          22.11        22.44
6.8          22.78        23.12
6.9          23.46        23.8
```

```
7.0           24.15          24.5
7.1           24.85          25.2
7.2           25.56          25.92
7.3           26.28          26.64
7.4           27.01          27.38
7.5           27.75          28.12
7.6           28.5           28.88
7.7           29.26          29.64
7.8           30.03          30.42
7.9           30.81          31.2
8.0           31.6           32.0
8.1           32.4           32.8
8.2           33.21          33.62
8.3           34.03          34.44
8.4           34.86          35.28
8.5           35.7           36.12
8.6           36.55          36.98
8.7           37.41          37.84
8.8           38.28          38.72
8.9           39.16          39.6
9.0           40.05          40.5
9.1           40.95          41.4
9.2           41.86          42.32
9.3           42.78          43.24
9.4           43.71          44.18
9.5           44.65          45.12
9.6           45.6           46.08
9.7           46.56          47.04
9.8           47.53          48.02
9.9           48.51          49.0
```

Not bad. Euler's method came pretty close to the exact solution. In fact, at a value of $x = 9.9$, Euler's method is still within 1% of the exact solution.

# Moving On Up with the Improved Euler's Method

As I explain in the earlier section "The fundamentals of the method," the recurrence relation for the Euler method is given by:

$$y_{n+1} = y_n + f(x_n, y_n)\, h$$

where the differential equation you're trying to solve is:

$$\frac{dy}{dx} = f(x,y)$$

This is a fairly simplistic method; it just extrapolates from the current point to the next point using the known slope. What if the actual solution varies faster than the Euler method takes into account?

Well, it turns out that an improved Euler method exists. This method, which is also called the Heun formula, can be more accurate. I explain everything you need to know in the following sections.

## Understanding the improvements

The standard Euler method assumes that the difference between $y_{n+1}$ and $y_n$ is:

$$\Delta y = f(x_n)\, h$$

For simplicity's sake, assume that $f(x, y)$ is only a function of $x$, $f(x)$.

This method doesn't take into account possible steep increases — or decreases — in the slope. A better solution would not only take the current slope, $f(x_n)$, into account, but it also would take the slope at the next point along the curve, $f(x_n + h)$, into account. You could then take the average of those two slopes, like this:

$$m = \frac{f(x_n) + f(x_n + h)}{2}$$

This is a better approximation than simply using the slope at $x_n$, which the standard Euler's method does. Now you can multiply the average slope by the interval length, $h$, to find the change in $y$:

$$\Delta y = \frac{h}{2}\left[f(x_n) + f(x_n + h)\right]$$

REMEMBER

So you can express the recurrence relation for the (new and) improved Euler's method like this:

$$y_{n+1} = y_n + \frac{h}{2}\left[f(x_n) + f(x_n + h)\right]$$

## Coming up with new code

In this section, I show you how to create a new program, e2.java, that will put the improved Euler method to work. The program will display the traditional Euler result, the improved Euler result, and the exact result. To illustrate the new code, I use the same equation from the earlier section "Using code to see the method in action" in the following sections.

### *Changes in the new code*

The code for the improved Euler's method is different from the one in the earlier section "Using code to see the method in action." Here, you create a new variable, yimp, in which you store the improved Euler results, and then you make these changes to the calculate method (the **bold** indicates changes in the code):

```
public void calculate()
{
  double x = x0;
  double y = y0;
  double yimp = y0;
  double k;

    System.out.println("x\t\tEuler\t\tImproved\t\tExact");

  for (int i = 1; i < n; i++){
    k = f(x, y);
    y - y + h * k;
    yimp = y + (f(x, y) + f(x + h, y))* (h/2);
    x = x + h;
    System.out.println(round(x) + "\t\t" + round(y) +
           "\t\t" + round(yimp) + "\t\t"
      + round(exact(x, 0)));
  }

}
```

Here's what the program looks like now:

```
public class e2
{

double x0 = 0.0;
double y0 = 0.0;
double h = 0.1;
double n = 100;

public e2()
{
}

public double f(double x, double y)
{
  return x;
}

public double exact(double x, double y)
{
  return x * x / 2;
}
```

```
public static void main(String [] argv)
{
  e2 de = new e2();
  de.calculate();
}

public void calculate()
{
  double x = x0;
  double y = y0;
  double yimp = y0;
  double k;

    System.out.println("x\t\tEuler\t\tImproved\t\tExact");

  for (int i = 1; i < n; i++){
    k = f(x, y);
    y = y + h * k;
    yimp = y + (f(x, y) + f(x + h, y))* (h/2);
    x = x + h;
    System.out.println(round(x) + "\t\t" + round(y) +
          "\t\t" + round(yimp) + "\t\t"
      + round(exact(x, 0)));
  }

}

public double round(double val)
{
  double divider = 100;
  val = val * divider;
  double temp = Math.round(val);
  return (double)temp / divider;
}
}
```

### *The results of the new code*

So how does the code work out? Take a look:

```
C:\>java e2
x               Euler           Improved        Exact
0.1             0.0             0.01            0.01
0.2             0.01            0.03            0.02
0.3             0.03            0.06            0.05
0.4             0.06            0.1             0.08
0.5             0.1             0.15            0.13
0.6             0.15            0.21            0.18
0.7             0.21            0.28            0.24
0.8             0.28            0.36            0.32
0.9             0.36            0.45            0.4
1.0             0.45            0.55            0.5
```

| | | | |
|---|---|---|---|
| 1.1 | 0.55 | 0.66 | 0.6 |
| 1.2 | 0.66 | 0.78 | 0.72 |
| 1.3 | 0.78 | 0.91 | 0.85 |
| 1.4 | 0.91 | 1.05 | 0.98 |
| 1.5 | 1.05 | 1.2 | 1.13 |
| 1.6 | 1.2 | 1.36 | 1.28 |
| 1.7 | 1.36 | 1.53 | 1.45 |
| 1.8 | 1.53 | 1.71 | 1.62 |
| 1.9 | 1.71 | 1.9 | 1.81 |
| 2.0 | 1.9 | 2.1 | 2.0 |
| 2.1 | 2.1 | 2.31 | 2.21 |
| 2.2 | 2.31 | 2.53 | 2.42 |
| 2.3 | 2.53 | 2.76 | 2.65 |
| 2.4 | 2.76 | 3.0 | 2.88 |
| 2.5 | 3.0 | 3.25 | 3.13 |
| 2.6 | 3.25 | 3.51 | 3.38 |
| 2.7 | 3.51 | 3.78 | 3.65 |
| 2.8 | 3.78 | 4.06 | 3.92 |
| 2.9 | 4.06 | 4.35 | 4.21 |
| 3.0 | 4.35 | 4.65 | 4.5 |
| 3.1 | 4.65 | 4.96 | 4.81 |
| 3.2 | 4.96 | 5.28 | 5.12 |
| 3.3 | 5.28 | 5.61 | 5.45 |
| 3.4 | 5.61 | 5.95 | 5.78 |
| 3.5 | 5.95 | 6.3 | 6.13 |
| 3.6 | 6.3 | 6.66 | 6.48 |
| 3.7 | 6.66 | 7.03 | 6.85 |
| 3.8 | 7.03 | 7.41 | 7.22 |
| 3.9 | 7.41 | 7.8 | 7.61 |
| 4.0 | 7.8 | 8.2 | 8.0 |
| 4.1 | 8.2 | 8.61 | 8.41 |
| 4.2 | 8.61 | 9.03 | 8.82 |
| 4.3 | 9.03 | 9.46 | 9.25 |
| 4.4 | 9.46 | 9.9 | 9.68 |
| 4.5 | 9.9 | 10.35 | 10.13 |
| 4.6 | 10.35 | 10.81 | 10.58 |
| 4.7 | 10.81 | 11.28 | 11.04 |
| 4.8 | 11.28 | 11.76 | 11.52 |
| 4.9 | 11.76 | 12.25 | 12.0 |
| 5.0 | 12.25 | 12.75 | 12.5 |
| 5.1 | 12.75 | 13.26 | 13.0 |
| 5.2 | 13.26 | 13.78 | 13.52 |
| 5.3 | 13.78 | 14.31 | 14.04 |
| 5.4 | 14.31 | 14.85 | 14.58 |
| 5.5 | 14.85 | 15.4 | 15.12 |
| 5.6 | 15.4 | 15.96 | 15.68 |
| 5.7 | 15.96 | 16.53 | 16.24 |
| 5.8 | 16.53 | 17.11 | 16.82 |
| 5.9 | 17.11 | 17.7 | 17.4 |
| 6.0 | 17.7 | 18.29 | 18.0 |
| 6.1 | 18.3 | 18.9 | 18.6 |
| 6.2 | 18.91 | 19.52 | 19.22 |

| | | | |
|---|---|---|---|
| 6.3 | 19.53 | 20.15 | 19.84 |
| 6.4 | 20.16 | 20.79 | 20.48 |
| 6.5 | 20.8 | 21.44 | 21.12 |
| 6.6 | 21.45 | 22.1 | 21.78 |
| 6.7 | 22.11 | 22.77 | 22.44 |
| 6.8 | 22.78 | 23.45 | 23.12 |
| 6.9 | 23.46 | 24.14 | 23.8 |
| 7.0 | 24.15 | 24.84 | 24.5 |
| 7.1 | 24.85 | 25.55 | 25.2 |
| 7.2 | 25.56 | 26.27 | 25.92 |
| 7.3 | 26.28 | 27.0 | 26.64 |
| 7.4 | 27.01 | 27.74 | 27.38 |
| 7.5 | 27.75 | 28.49 | 28.12 |
| 7.6 | 28.5 | 29.25 | 28.88 |
| 7.7 | 29.26 | 30.02 | 29.64 |
| 7.8 | 30.03 | 30.8 | 30.42 |
| 7.9 | 30.81 | 31.59 | 31.2 |
| 8.0 | 31.6 | 32.39 | 32.0 |
| 8.1 | 32.4 | 33.2 | 32.8 |
| 8.2 | 33.21 | 34.02 | 33.62 |
| 8.3 | 34.03 | 34.85 | 34.44 |
| 8.4 | 34.86 | 35.69 | 35.28 |
| 8.5 | 35.7 | 36.54 | 36.12 |
| 8.6 | 36.55 | 37.4 | 36.98 |
| 8.7 | 37.41 | 38.27 | 37.84 |
| 8.8 | 38.28 | 39.15 | 38.72 |
| 8.9 | 39.16 | 40.04 | 39.6 |
| 9.0 | 40.05 | 40.94 | 40.5 |
| 9.1 | 40.95 | 41.85 | 41.4 |
| 9.2 | 41.86 | 42.77 | 42.32 |
| 9.3 | 42.78 | 43.7 | 43.24 |
| 9.4 | 43.71 | 44.64 | 44.18 |
| 9.5 | 44.65 | 45.59 | 45.12 |
| 9.6 | 45.6 | 46.55 | 46.08 |
| 9.7 | 46.56 | 47.52 | 47.04 |
| 9.8 | 47.53 | 48.5 | 48.02 |
| 9.9 | 48.51 | 49.49 | 49.0 |

As you can see, Euler's method and the improved Euler's method come out about the same here. They're both off by 1% at $x = 9.9$.

## Plugging a steep slope into the new code

REMEMBER

The improvement in the improved Euler's method isn't often visible until you have steep or quickly varying slopes. For example, take a look at this differential equation:

$$\frac{dy}{dx} = x^6$$

where $y(0) = 0$.

Here's the exact solution (I'll spare you the calculations so you can get to the coding faster):

$$y = \frac{x^7}{7}$$

Okay, so if the improved Euler method is truly improved, you should be able to see difference in this example, where the slope changes faster than in the differential equation in the previous section. Keep reading to find out what happens.

### Checking out the results

As you can see, when you use the steeply sloping differential equation, there's a difference between the Euler method and the improved Euler method:

| x | Euler | Improved | Exact |
|---|---|---|---|
| | . | | |
| | . | | |
| | . | | |
| 6.0 | 37696.93 | 42138.76 | 39990.86 |
| 6.1 | 42362.53 | 47271.35 | 44896.33 |
| 6.2 | 47514.57 | 52930.6 | 50308.78 |
| 6.3 | 53194.59 | 59160.78 | 56271.15 |
| 6.4 | 59446.94 | 66009.09 | 62829.24 |
| 6.5 | 66310.89 | 73525.81 | 70031.83 |
| 6.6 | 73860.78 | 81764.42 | 77930.87 |
| 6.7 | 82126.18 | 90781.79 | 86581.59 |
| 6.8 | 91172.01 | 100638.31 | 96042.7 |
| 6.9 | 101058.76 | 111398.05 | 106376.48 |
| 7.0 | 111850.58 | 123128.94 | 117649.0 |
| 7.1 | 123615.48 | 135902.94 | 129930.29 |
| 7.2 | 136425.51 | 149796.23 | 143294.47 |
| 7.3 | 150356.91 | 164889.33 | 157819.98 |
| 7.4 | 165490.34 | 181267.37 | 173589.72 |
| 7.5 | 181910.99 | 199020.24 | 190691.27 |
| 7.6 | 199708.84 | 218242.76 | 209217.07 |
| 7.7 | 218978.83 | 239034.95 | 229264.62 |
| 7.8 | 239821.07 | 261502.17 | 250936.7 |
| 7.9 | 263341.03 | 285755.30 | 274341.56 |
| 8.0 | 286649.77 | 311911.35 | 299593.14 |
| 8.1 | 312864.17 | 340092.85 | 326811.32 |
| 8.2 | 341107.13 | 370428.94 | 356122.1 |
| 8.3 | 371507.8 | 403055.15 | 387657.87 |
| 8.4 | 404201.83 | 438113.75 | 421557.64 |
| 8.5 | 439331.64 | 475754.01 | 457967.27 |
| 8.6 | 477046.59 | 516132.42 | 497039.75 |
| 8.7 | 517503.31 | 559412.98 | 538935.42 |
| 8.8 | 560865.93 | 605767.45 | 583822.28 |
| 8.9 | 607306.34 | 655375.61 | 631876.21 |
| 9.0 | 657004.47 | 708425.58 | 683281.29 |
| 9.1 | 710148.57 | 765114.08 | 738230.03 |
| 9.2 | 766935.49 | 825646.71 | 796923.72 |

```
9.3     827570.99       890238.25       859572.67
9.4     892270.01       959113.01       926396.56
9.5     961256.99       1032505.07      997624.71
9.6     1034766.18      1110658.66      1073496.4
9.7     1113041.96      1193828.45      1154261.21
9.8     1196339.16      1282279.88      1240179.33
9.9     1284923.4       1376289.52      1331521.93
```

Even though there is indeed a difference here (at $x = 9.9$, the Euler method is off by 3.5% and the improved Euler method is off by 3.4%), the difference isn't huge.

### Increasing your accuracy with a decreased step size

TIP

One way to increase your accuracy is to decrease your step size. For instance, decrease the step size from $h = 0.1$ to $h = 0.01$. Also increase the number of steps from $n = 100$ to $n = 1{,}000$. Here are the changes to make in the code in e2.java:

```
public class e2
{

double x0 = 0.0;
double y0 = 0.0;
double h = 0.01;
double n = 1000;
          .
          .
          .
```

And here's what the results look like:

```
x       Euler           Improved              Exact
         .
         .
         .
9.0     680627.03       685923.78             683281.29
9.01    685941.44       691273.62             688613.44
9.02    691291.38       696659.18             693981.23
9.03    696677.04       702080.67             699384.84
9.04    702098.63       707538.28             704824.47
9.05    707556.34       713032.22             710300.33
9.06    713050.38       718562.68             715812.61
9.07    718580.94       724129.87             721361.52
9.08    724148.23       729733.98             726947.25
9.09    729752.45       735375.23             732570.02
9.1     735393.8        741053.82             738230.03
9.11    741072.49       746769.96             743927.48
9.12    746788.73       752523.84             749662.57
9.13    752542.72       758315.69             755435.52
```

| | | | |
|---|---|---|---|
| 9.14 | 758334.67 | 764145.7 | 761246.54 |
| 9.15 | 764164.78 | 770014.09 | 767095.82 |
| 9.16 | 770033.28 | 775921.06 | 772983.59 |
| 9.17 | 775940.36 | 781866.84 | 778910.05 |
| 9.18 | 781886.24 | 787851.62 | 784875.42 |
| 9.19 | 787871.12 | 793875.62 | 790879.9 |
| 9.2 | 793895.24 | 799939.07 | 796923.72 |
| 9.21 | 799958.79 | 806042.16 | 803007.07 |
| 9.22 | 806061.99 | 812185.13 | 809130.19 |
| 9.23 | 812205.06 | 818368.17 | 815293.29 |
| 9.24 | 818388.22 | 824591.52 | 821496.58 |
| 9.25 | 824611.67 | 830855.39 | 827740.28 |
| 9.26 | 830875.66 | 837160.01 | 834024.61 |
| 9.27 | 837180.38 | 843505.58 | 840349.8 |
| 9.28 | 843526.06 | 849892.34 | 846716.05 |
| 9.29 | 849912.93 | 856320.5 | 853123.6 |
| 9.3 | 856341.2 | 862790.29 | 859572.67 |
| 9.31 | 862811.1 | 869301.93 | 866063.48 |
| 9.32 | 869322.86 | 875855.65 | 872596.26 |
| 9.33 | 875876.69 | 882451.68 | 879171.23 |
| 9.34 | 882472.83 | 889090.24 | 885788.61 |
| 9.35 | 889111.5 | 895771.56 | 892448.65 |
| 9.36 | 895792.94 | 902495.87 | 899151.56 |
| 9.37 | 902517.36 | 909263.4 | 905897.57 |
| 9.38 | 909285.01 | 916074.38 | 912686.92 |
| 9.39 | 916096.1 | 922929.04 | 919519.84 |
| 9.4 | 922950.88 | 929827.62 | 926396.56 |
| 9.41 | 929849.58 | 936770.36 | 933317.32 |
| 9.42 | 936792.43 | 943757.47 | 940282.34 |
| 9.43 | 943779.67 | 950789.21 | 947291.87 |
| 9.44 | 950811.53 | 957865.82 | 954346.14 |
| 9.45 | 957000.25 | 964007.51 | 961445.39 |
| 9.46 | 965010.06 | 972154.55 | 968589.85 |
| 9.47 | 972177.22 | 979367.16 | 975779.78 |
| 9.48 | 979389.95 | 986625.59 | 983015.4 |
| 9.49 | 986648.5 | 993930.09 | 990296.96 |
| 9.5 | 993953.12 | 1001280.88 | 997624.71 |
| 9.51 | 1001304.04 | 1008678.23 | 1004998.88 |
| 9.52 | 1008701.51 | 1016122.37 | 1012419.73 |
| 9.53 | 1016145.77 | 1023613.55 | 1019887.49 |
| 9.54 | 1023637.07 | 1031152.02 | 1027402.42 |
| 9.55 | 1031175.66 | 1038738.02 | 1034964.76 |
| 9.56 | 1038761.79 | 1046371.82 | 1042574.76 |
| 9.57 | 1046395.71 | 1054053.64 | 1050232.67 |
| 9.58 | 1054077.66 | 1061783.76 | 1057938.75 |
| 9.59 | 1061807.9 | 1069562.41 | 1065693.24 |
| 9.6 | 1069586.69 | 1077389.87 | 1073496.4 |
| 9.61 | 1077414.26 | 1085266.37 | 1081348.48 |
| 9.62 | 1085290.89 | 1093192.17 | 1089249.74 |
| 9.63 | 1093216.82 | 1101167.54 | 1097200.43 |
| 9.64 | 1101192.32 | 1109192.73 | 1105200.82 |

| | | | |
|---|---|---|---|
| 9.65 | 1109217.64 | 1117268.0 | 1113251.15 |
| 9.66 | 1117293.04 | 1125393.6 | 1121351.7 |
| 9.67 | 1125418.77 | 1133569.81 | 1129502.71 |
| 9.68 | 1133595.11 | 1141796.88 | 1137704.46 |
| 9.69 | 1141822.31 | 1150075.08 | 1145957.2 |
| 9.7 | 1150100.64 | 1158404.66 | 1154261.21 |
| 9.71 | 1158430.36 | 1166785.91 | 1162616.73 |
| 9.72 | 1166811.74 | 1175219.08 | 1171024.05 |
| 9.73 | 1175245.04 | 1183704.43 | 1179483.42 |
| 9.74 | 1183730.53 | 1192242.25 | 1187995.12 |
| 9.75 | 1192268.48 | 1200832.8 | 1196559.42 |
| 9.76 | 1200859.17 | 1209476.35 | 1205176.58 |
| 9.77 | 1209502.85 | 1218173.17 | 1213846.88 |
| 9.78 | 1218199.81 | 1226923.54 | 1222570.59 |
| 9.79 | 1226950.31 | 1235727.73 | 1231347.98 |
| 9.8 | 1235754.64 | 1244586.02 | 1240179.33 |
| 9.81 | 1244613.06 | 1253498.68 | 1249064.92 |
| 9.82 | 1253525.86 | 1262465.99 | 1258005.02 |
| 9.83 | 1262493.31 | 1271488.23 | 1266999.91 |
| 9.84 | 1271515.69 | 1280565.68 | 1276049.87 |
| 9.85 | 1280593.28 | 1289698.62 | 1285155.19 |
| 9.86 | 1289726.36 | 1298887.33 | 1294316.13 |
| 9.87 | 1298915.22 | 1308132.11 | 1303533.0 |
| 9.88 | 1308160.14 | 1317433.22 | 1312806.06 |
| 9.89 | 1317461.39 | 1326790.97 | 1322135.61 |
| 9.9 | 1326819.28 | 1336205.62 | 1331521.93 |

When using the smaller step size, the Euler method and the improved Euler method are only off by about 0.35% at $x = 9.9$. Much better.

# Adding Even More Precision with the Runge-Kutta Method

If you don't want to use either of the Euler methods, you're in luck. There's another numerical method that you can use to solve differential equations: the Runge-Kutta method. This method gives excellent results that are even more accurate than the Euler or improved Euler methods.

## The method's recurrence relation

REMEMBER

In the Runge-Kutta method, the recurrence relation is a weighted average of terms:

$$y_{n+1} = y_n + \frac{h}{6}[c_1 + c_2 + c_3 + c_4]$$

## The story of Runge and Kutta

Carl David Tolmé Runge was a German mathematician. He was born in Havana, Cuba, on August 30, 1856, where his father was the Danish consul. The family later moved to Bremen, Germany. Runge died on January 3, 1927.

Martin Wilhelm Kutta was also a German mathematician. He was born in Pitschen, Germany (which today is part of Poland), on November 3, 1867. He went to the University of Breslau and continued to work in Munich. He also spent a year at the University of Cambridge. He became a professor in Stuttgart, Germany, in 1911. Kutta died on December 25, 1944.

In 1901, Runge and Kutta co-developed the Runge-Kutta method, which is the powerful method that's used to solve ordinary differential equations numerically.

where:

$$c_1 = f(x_n\ y_n)$$

$$c_2 = f\left(x_n + \frac{h}{2},\ y_n + c_1\frac{h}{2}\right)$$

$$c_3 = f\left(x_n + \frac{h}{2},\ y_n + c_2\frac{h}{2}\right)$$

$$c_4 = f(x_n + h,\ y_n + c_3 h)$$

REMEMBER

This method can easily be adapted to the code you use earlier in the chapter, because $f(x, y)$ doesn't depend on $y$ — it's just $f(x)$. In this case, the recurrence relation becomes:

$$y_{n+1} = y_n + \frac{h}{6}\left[f(x_n) + f\left(x_n + \frac{h}{2}\right) + f\left(x_n + \frac{h}{2}\right) + f(x_n + h)\right]$$

# *Working with the method in code*

Here, you put the Runge-Kutta technique to work, solving the same differential equation that you saw earlier in the chapter:

$$\frac{dy}{dx} = x$$

### *Inputting the code*

The recurrence relation of the Runge-Kutta method looks easy enough to implement in a new Java program, e3.java. This new program will put the Runge-Kutta method to work. The new code is shown in **bold.**

```
public class e3
{

double x0 = 0.0;
double y0 = 0.0;
double h = 0.1;
double n = 100;

public e3()
{
}

public double f(double x, double y)
{
  return x;
}

public double exact(double x, double y)
{
  return x * x / 2;
}

public static void main(String [] argv)
{
  e3 de = new e3();
  de.calculate();
}

public void calculate()
{
  double x = x0;
  double y = y0;
  double yrk = y0;
  double k;

    System.out.println("x\t\tEuler\t\tRunge-
          Kutta\t\tExact");

  for (int i = 1; i < n; i++){
    k = f(x, y);
    y = y + h * k;
    yrk = y + (f(x, y) + f(x + h/2, y)/2 + f(x + h/2, y)/2
          + f(x + h, y))* (h/6);
    x = x + h;
    System.out.println(round(x) + "\t\t" + round(y) +
          "\t\t" + round(yrk) + "\t\t"
      + round(exact(x, 0)));
  }

}

public double round(double val)
```

```
{
  double divider = 100;
  val = val * divider;
  double temp = Math.round(val);
  return (double)temp / divider;
}
}
```

### *Examining the results*

Here are the results of e3.java:

```
C:\>java e3
x                Euler            Runge-Kutta      Exact
0.1              0.0              0.0              0.01
0.2              0.01             0.02             0.02
0.3              0.03             0.04             0.05
0.4              0.06             0.08             0.08
0.5              0.1              0.12             0.13
0.6              0.15             0.18             0.18
0.7              0.21             0.24             0.24
0.8              0.28             0.32             0.32
0.9              0.36             0.4              0.4
1.0              0.45             0.5              0.5
1.1              0.55             0.6              0.6
1.2              0.66             0.72             0.72
1.3              0.78             0.84             [illegible]
1.4              0.91             0.98             0.98
1.5              1.05             1.12             1.13
1.6              1.2              1.28             1.28
1.7              1.36             1.44             1.45
1.8              1.53             1.62             1.62
1.9              1.71             1.8              1.81
2.0              1.9              2.0              2.0
2.1              2.1              2.2              2.21
2.2              2.31             2.42             2.42
2.3              2.53             2.64             2.65
2.4              2.76             2.88             2.88
2.5              3.0              3.12             3.13
2.6              3.25             3.38             3.38
2.7              3.51             3.64             3.65
2.8              3.78             3.92             3.92
2.9              4.06             4.2              4.21
3.0              4.35             4.5              4.5
3.1              4.65             4.8              4.81
3.2              4.96             5.12             5.12
3.3              5.28             5.44             5.45
3.4              5.61             5.78             5.78
3.5              5.95             6.12             6.13
3.6              6.3              6.48             6.48
3.7              6.66             6.84             6.85
3.8              7.03             7.22             7.22
```

| | | | |
|---|---|---|---|
| 3.9 | 7.41 | 7.6 | 7.61 |
| 4.0 | 7.8 | 8.0 | 8.0 |
| 4.1 | 8.2 | 8.4 | 8.41 |
| 4.2 | 8.61 | 8.82 | 8.82 |
| 4.3 | 9.03 | 9.24 | 9.25 |
| 4.4 | 9.46 | 9.68 | 9.68 |
| 4.5 | 9.9 | 10.12 | 10.13 |
| 4.6 | 10.35 | 10.58 | 10.58 |
| 4.7 | 10.81 | 11.04 | 11.04 |
| 4.8 | 11.28 | 11.52 | 11.52 |
| 4.9 | 11.76 | 12.0 | 12.0 |
| 5.0 | 12.25 | 12.5 | 12.5 |
| 5.1 | 12.75 | 13.0 | 13.0 |
| 5.2 | 13.26 | 13.52 | 13.52 |
| 5.3 | 13.78 | 14.04 | 14.04 |
| 5.4 | 14.31 | 14.58 | 14.58 |
| 5.5 | 14.85 | 15.12 | 15.12 |
| 5.6 | 15.4 | 15.68 | 15.68 |
| 5.7 | 15.96 | 16.24 | 16.24 |
| 5.8 | 16.53 | 16.82 | 16.82 |
| 5.9 | 17.11 | 17.4 | 17.4 |
| 6.0 | 17.7 | 18.0 | 18.0 |
| 6.1 | 18.3 | 18.6 | 18.6 |
| 6.2 | 18.91 | 19.22 | 19.22 |
| 6.3 | 19.53 | 19.84 | 19.84 |
| 6.4 | 20.16 | 20.48 | 20.48 |
| 6.5 | 20.8 | 21.12 | 21.12 |
| 6.6 | 21.45 | 21.78 | 21.78 |
| 6.7 | 22.11 | 22.44 | 22.44 |
| 6.8 | 22.78 | 23.12 | 23.12 |
| 6.9 | 23.46 | 23.8 | 23.8 |
| 7.0 | 24.15 | 24.5 | 24.5 |
| 7.1 | 24.85 | 25.2 | 25.2 |
| 7.2 | 25.56 | 25.92 | 25.92 |
| 7.3 | 26.28 | 26.64 | 26.64 |
| 7.4 | 27.01 | 27.38 | 27.38 |
| 7.5 | 27.75 | 28.12 | 28.12 |
| 7.6 | 28.5 | 28.88 | 28.88 |
| 7.7 | 29.26 | 29.64 | 29.64 |
| 7.8 | 30.03 | 30.42 | 30.42 |
| 7.9 | 30.81 | 31.2 | 31.2 |
| 8.0 | 31.6 | 32.0 | 32.0 |
| 8.1 | 32.4 | 32.8 | 32.8 |
| 8.2 | 33.21 | 33.62 | 33.62 |
| 8.3 | 34.03 | 34.44 | 34.44 |
| 8.4 | 34.86 | 35.28 | 35.28 |
| 8.5 | 35.7 | 36.12 | 36.12 |
| 8.6 | 36.55 | 36.98 | 36.98 |
| 8.7 | 37.41 | 37.84 | 37.84 |
| 8.8 | 38.28 | 38.72 | 38.72 |

| | | | |
|---|---|---|---|
| 8.9 | 39.16 | 39.6 | 39.6 |
| 9.0 | 40.05 | 40.5 | 40.5 |
| 9.1 | 40.95 | 41.4 | 41.4 |
| 9.2 | 41.86 | 42.32 | 42.32 |
| 9.3 | 42.78 | 43.24 | 43.24 |
| 9.4 | 43.71 | 44.18 | 44.18 |
| 9.5 | 44.65 | 45.12 | 45.12 |
| 9.6 | 45.6 | 46.08 | 46.08 |
| 9.7 | 46.56 | 47.04 | 47.04 |
| 9.8 | 47.53 | 48.02 | 48.02 |
| 9.9 | 48.51 | 49.0 | 49.0 |

When you look at the results, you can see that the Runge-Kutta method beat the Euler method hands down. In fact, to two decimal places, the Runge-Kutta method nearly always nailed the correct answer. Very cool.

# Part IV
# The Part of Tens

The 5th Wave By Rich Tennant

"I looked over your equations, Mrs. Dundt. Your concavity and inflection points are clean and there's nothing wrong with your velocity and acceleration. It might be your differentiation, but I won't be able to look at it until Thursday."

## *In this part . . .*

Every *For Dummies* book features a Part of Tens, and who am I to break with tradition? This part is full of cool stuff. You discover the ten top differential equation online tutorials as well as the ten top online tools for solving differential equations. If you want to get more info on a specific aspect of differential equations, try the tutorials. If you want to work with challenging differential equations and need a little help, take a look at the tools available online.

# Chapter 14

# Ten Super-Helpful Online Differential Equation Tutorials

### In This Chapter

- Understanding different aspects of differential equations with tutorials
- Checking out helpful notes and videos

Even if you think of yourself as a math whiz, you still may find yourself scratching your head at certain aspects of differential equations. Fear not: A number of differential equation tutorials are available on the Internet — and this chapter lists ten of my favorites. Each site deals with various aspects of differential equations.

## AnalyzeMath.com's Introduction to Differential Equations

If you're new to differential equations, you may be looking for a brief overview of the topic. The Web site AnalyzeMath.com offers just that. If you visit `www.analyzemath.com/calculus/Differential_Equations/introduction.html`, you can read an introductory tutorial on differential equations. The site is run by Abdelkader Dendane, PhD, a lecturer in mathematics at United Arab Emirates University.

## Harvey Mudd College Mathematics Online Tutorial

If you're looking for a tutorial that gives good coverage of solving first order ordinary differential equations, be sure to check out the Harvey Mudd College Mathematics Online Tutorial at `www.math.hmc.edu/calculus/tutorials/odes`.

## John Appleby's Introduction to Differential Equations

John Appleby, a professor in the School of Mathematical Sciences at Dublin City University in Ireland, maintains a helpful tutorial at `webpages.dcu.ie/~applebyj/ms225/ms225.html`. Be sure to click the links under the Tutorial Sheets and the Supplementary Notes headers. On his site, Appleby covers separable and homogeneous equations, first order linear differential equations, second order linear differential equations, variation of parameters, and more.

## Kardi Teknomo's Page

Dr. Kardi Teknomo is a research fellow at Human Centered Mobility Technologies in Arsenal Research (which is located in Austria). Teknomo's tutorial focuses on solving differential equations using numerical methods. You can find this great tutorial at `people.revoledu.com/kardi/tutorial/ODE/index.html`.

## Martin J. Osborne's Differential Equation Tutorial

Martin J. Osborne, a professor of economics at the University of Toronto, has created a superb, multipart differential equation tutorial at `www.economics.utoronto.ca/osborne/MathTutorial/IDEF.HTM`.

# Midnight Tutor's Video Tutorial

If you're looking for something fun and different, check out the Midnight Tutor's Video Tutorial, which is a video tutorial solving $y' - y\cos(t) = \cos(t)$ using separation of variables. To view the video, visit `www.midnighttutor.com/de_xprime-xcost-xcost.html`.

# The Ohio State University Physics Department's Introduction to Differential Equations

If you struggle with homogeneous and nonhomogeneous linear differential equations, there's still hope! Check out the Ohio State University Physics Department tutorial on these types of differential equations. You can find the site at `www.physics.ohio-state.edu/~physedu/mapletutorial/tutorials/diff_eqs/intro.html`.

# Paul's Online Math Notes

Paul's Online Math Notes is an extensive set of online explanations of differential equations. You can visit this site at `tutorial.math.lamar.edu/Classes/DE/DE.aspx`. (Who's Paul? He's Paul Dawkins, who teaches at Lamar University in Beaumont, Texas.)

# S.O.S. Math

S.O.S. Math (`www.sosmath.com/diffeq/diffeq.html`) is an all-purpose resource for all of mathematics, but it has an entire section on differential equations. S.O.S Math is a great, multipart tutorial — it's as complete a tutorial as you'll find online.

## University of Surrey Tutorial

England's University of Surrey provides an excellent tutorial on first and second order differential equations. You can view this site at `www.maths.surrey.ac.uk/explore/vithyaspages`. Vithya Nanthakumaar, who studies at the University of Surrey, created the site.

# Chapter 15

# Ten Really Cool Online Differential Equation Solving Tools

*In This Chapter*

- Solving equations with handy online tools
- Plotting direction fields and graphing functions

Suppose you're working on a particularly hairy differential equation that requires some busywork. Do you need to sit and do pages of calculations? Heck no! That's what differential equation solving tools are for! And tons are available. Some programs cost money — and sometimes, they cost a great deal of money. However, a bunch of good tools also are available for free online. This chapter has a sampling of those great freebies.

## AnalyzeMath.com's Runge-Kutta Method Applet

The AnalyzeMath.com Runge-Kutta Applet solves a number of representative differential equations using the Runge-Kutta method that I explain in Chapter 13. You can find this applet at `www.analyzemath.com/calculus/RungeKutta/RungeKutta.html`. A couple of cool features of this applet: You can increase the number of points used in the calculation, and you can zoom in and out.

## Coolmath.com's Graphing Calculator

The Coolmath.com graphing calculator is a fabulous online graphing applet that draws functions for you. You can find this calculator at `www.coolmath.com/graphit/index.html`.

## Direction Field Plotter

Want to see what the direction field for a differential equation's solution looks like? You're in luck! You can find a great direction field plotter at `www.math.ubc.ca/~israel/applet/dfplotter/dfplotter.html`. (Flip to Chapter 1 for an introduction to direction fields.)

## An Equation Solver from QuickMath Automatic Math Solutions

TIP

The solver from QuickMath Automatic Math Solutions is an equation solver, not a differential equation solver. But it's still a great tool when you need to factor an equation, such as a characteristic equation, to find the roots. Here's the Web site where you can find the solver: `www.hostsrv.com/webmab/app1/MSP/quickmath/02/pageGenerate?site=quickmath&s1=equations&s2=solve&s3=basic`.

## First Order Differential Equation Solver

Here's a cool one: the First Order Differential Equation Solver at `www.cs.gordon.edu/~senning/desolver/index.html`. With this solver you can use the Euler, Improved Euler, and Runge-Kutta methods to solve differential equations. Just select the method you want to use from the drop-down box, enter your equation, and then click the Submit button. You couldn't find a niftier math tool.

## GCalc Online Graphing Calculator

If you're working on a tough problem and don't have a graphing calculator handy, you can find a good graphing calculator applet at `www.calculator.com/calcs/GCalc.html`. Just enter the function you want to graph and press Enter on your keyboard. It doesn't get easier than that!

# JavaView Ode Solver

The JavaView Ode Solver at `www.javaview.de/services/odeSolver/index.html` is a numerical differential equation solver that uses the Runge-Kutta method in Chapter 13. ("Ode" stands for "ordinary differential equation.")

# Math @ CowPi's System Solver

The Math @ CowPi's System Solver lets you solve simultaneous equations — from $2 \times 2$ to $5 \times 5$. Just fill in all the blanks, and presto! You have your answer. This great tool is available at `math.cowpi.com/systemsolver`.

This tool also can be useful when you break a higher-order differential equation into a system of lower-order ones.

# A Matrix Inverter from QuickMath Automatic Math Solutions

When working with systems of differential equations, you work with equations like $\mathbf{Ax} = \mathbf{b}$. The solution to an equation like this is $\mathbf{x} = \mathbf{A}^{-1}\mathbf{b}$, where $\mathbf{A}^{-1}$ is the inverse of the matrix $\mathbf{A}$. Don't feel like solving all that? Here's a tool that finds the inverse of matrices for you in a snap: `www.hostsrv.com/webmab/app1/MSP/quickmath/02/pageGenerate?site=quickmath&s1=matrices&s2=inverse&s3=basic`. All you have to do is enter the matrix and click the Inverse button.

# Visual Differential Equation Solving Applet

The Visual Differential Equation Solving Applet at `www.falstad.com/diffeq` (which runs in browsers) lets you solve some common differential equations and adjust numeric parameters.

# Index

## • E •

## • F •



## • G •

## • H •



## • I •

## • J •



## • K •



## • L •

## • M •

# • N •



# • O •



# • P •

## • Q •



## • R •

## • S •

## • T •



## • U •

## • V •



## • W •



## • X •

# Notes

# Notes

## USINESS, CAREERS & PERSONAL FINANCE

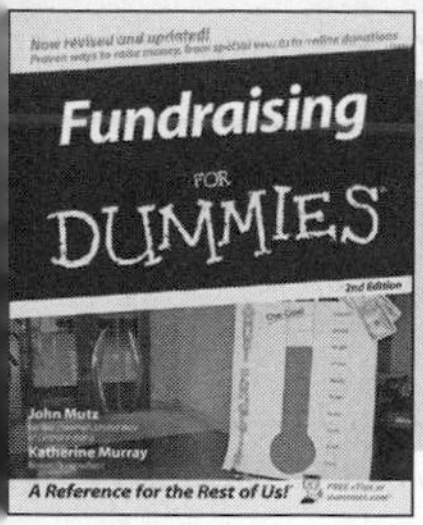


0-7645-9847-3

0-7645-2431-3

**Also available:**

- Business Plans Kit For Dummies 0-7645-9794-9
- Economics For Dummies 0-7645-5726-2
- Grant Writing For Dummies 0-7645-8416-2
- Home Buying For Dummies 0-7645-5331-3
- Managing For Dummies 0-7645-1771-6
- Marketing For Dummies 0-7645-5600-2
- Personal Finance For Dummies 0-7645-2590-5*
- Resumes For Dummies 0-7645-5471-9
- Selling For Dummies 0-7645-5363-1
- Six Sigma For Dummies 0-7645-6798-5
- Small Business Kit For Dummies 0-7645-5984-2
- Starting an eBay Business For Dummies 0-7645-6924-4
- Your Dream Career For Dummies 0-7645-9795-7

## OME & BUSINESS COMPUTER BASICS

0-470-05432-8

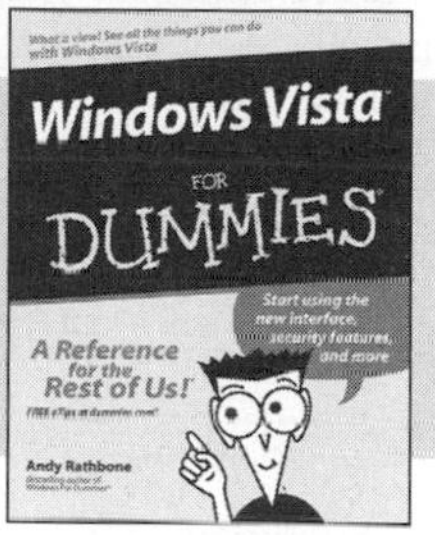


0-471-75421-8

**Also available:**

- Cleaning Windows Vista For Dummies 0-471-78293-9
- Excel 2007 For Dummies 0-470-03737-7
- Mac OS X Tiger For Dummies 0-7645-7675-5
- MacBook For Dummies 0-470-04859-X
- Macs For Dummies 0-470-04849-2
- Office 2007 For Dummies 0-470-00923-3
- Outlook 2007 For Dummies 0-470-03830-6
- PCs For Dummies 0-7645-8958-X
- Salesforce.com For Dummies 0-470-04893-X
- Upgrading & Fixing Laptops For Dummies 0-7645-8959-8
- Word 2007 For Dummies 0-470-03658-3
- Quicken 2007 For Dummies 0-470-04600-7

## OOD, HOME, GARDEN, HOBBIES, MUSIC & PETS

0-7645-8404-9

0-7645-9904-6

**Also available:**

- Candy Making For Dummies 0-7645-9734-5
- Card Games For Dummies 0-7645-9910-0
- Crocheting For Dummies 0-7645-4151-X
- Dog Training For Dummies 0-7645-8418-0
- Healthy Carb Cookbook For Dummies 0-7645-8476-6
- Home Maintenance For Dummies 0-7645-5215-5
- Horses For Dummies 0-7645-9797-3
- Jewelry Making & Beading For Dummies 0-7645-2571-9
- Orchids For Dummies 0-7645-6759-4
- Puppies For Dummies 0-7645-5255-4
- Rock Guitar For Dummies 0-7645-5356-9
- Sewing For Dummies 0-7645-6847-7
- Singing For Dummies 0-7645-2475-5

## ITERNET & DIGITAL MEDIA

0-470-04529-9

0-470-04894-8

**Also available:**

- Blogging For Dummies 0-471-77084-1
- Digital Photography For Dummies 0-7645-9802-3
- Digital Photography All-in-One Desk Reference For Dummies 0-470-03743-1
- Digital SLR Cameras and Photography For Dummies 0-7645-9803-1
- eBay Business All-in-One Desk Reference For Dummies 0-7645-8438-3
- HDTV For Dummies 0-470-09673-X
- Home Entertainment PCs For Dummies 0-470-05523-5
- MySpace For Dummies 0-470-09529-6
- Search Engine Optimization For Dummies 0-471-97998-8
- Skype For Dummies 0-470-04891-3
- The Internet For Dummies 0-7645-8996-2
- Wiring Your Digital Home For Dummies 0-471-91830-X

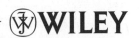

**eparate Canadian edition also available**
**eparate U.K. edition also available**

ilable wherever books are sold. For more information or to order direct: U.S. customers visit www.dummies.com or call 1-877-762-2974. . customers visit www.wileyeurope.com or call 0800 243407. Canadian customers visit www.wiley.ca or call 1-800-567-4797.

WILEY

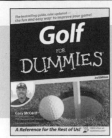

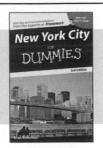

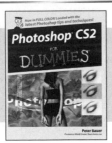

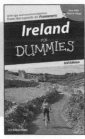

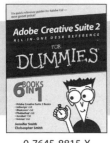

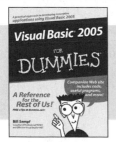

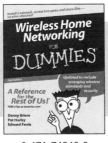

## ALTH & SELF-HELP

0-7645-8450-2

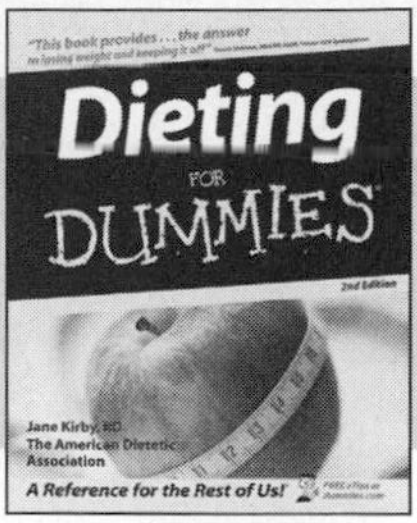


0-7645-4149-8

**Also available:**

- Bipolar Disorder For Dummies 0-7645-8451-0
- Chemotherapy and Radiation For Dummies 0-7645-7832-4
- Controlling Cholesterol For Dummies 0-7645-5440-9
- Diabetes For Dummies 0-7645-6820-5* †
- Divorce For Dummies 0-7645-8417-0 †
- Fibromyalgia For Dummies 0-7645-5441-7
- Low-Calorie Dieting For Dummies 0-7645-9905-4
- Meditation For Dummies 0-471-77774-9
- Osteoporosis For Dummies 0-7645-7621-6
- Overcoming Anxiety For Dummies 0-7645-5447-6
- Reiki For Dummies 0-7645-9907-0
- Stress Management For Dummies 0-7645-5144-2

## UCATION, HISTORY, REFERENCE & TEST PREPARATION

0-7645-8381-6

0-7645-9554-7

**Also available:**

- The ACT For Dummies 0-7645-9652-7
- Algebra For Dummies 0-7645-5325-9
- Algebra Workbook For Dummies 0-7645-8467-7
- Astronomy For Dummies 0-7645-8465-0
- Calculus For Dummies 0-7645-2498-4
- Chemistry For Dummies 0-7645-5430-1
- Forensics For Dummies 0-7645-5580-4
- Freemasons For Dummies 0-7645-9796-5
- French For Dummies 0-7645-5193-0
- Geometry For Dummies 0-7645-5324-0
- Organic Chemistry I For Dummies 0-7645-6902-3
- The SAT I For Dummies 0-7645-7193-1
- Spanish For Dummies 0-7645-5194-9
- Statistics For Dummies 0-7645-5423-9

# Get smart @ dummies.com®

- **Find a full list of Dummies titles**
- **Look into loads of FREE on-site articles**
- **Sign up for FREE eTips e-mailed to you weekly**
- **See what other products carry the Dummies name**
- **Shop directly from the Dummies bookstore**
- **Enter to win new prizes every month!**

**)arate Canadian edition also available**
**)arate U.K. edition also available**

able wherever books are sold. For more information or to order direct: U.S. customers visit www.dummies.com or call 1-877-762-2974. :ustomers visit www.wileyeurope.com or call 0800 243407. Canadian customers visit www.wiley.ca or call 1-800-567-4797.